Real-Time Embedded Systems

Real-Time Embedded Systems

Special Issue Editors

Christos Koulamas
Mihai T. Lazarescu

MDPI • Basel • Beijing • Wuhan • Barcelona • Belgrade

Special Issue Editors
Christos Koulamas
Industrial Systems Institute
Greece

Mihai T. Lazarescu
Politecnico di Torino
Italy

Editorial Office
MDPI
St. Alban-Anlage 66
4052 Basel, Switzerland

This is a reprint of articles from the Special Issue published online in the open access journal *Electronics* (ISSN 2079-9292) from 2017 to 2018 (available at: https://www.mdpi.com/journal/electronics/special_issues/embedded_systems)

For citation purposes, cite each article independently as indicated on the article page online and as indicated below:

LastName, A.A.; LastName, B.B.; LastName, C.C. Article Title. *Journal Name* **Year**, *Article Number*, Page Range.

ISBN 978-3-03897-509-0 (Pbk)
ISBN 978-3-03897-510-6 (PDF)

Contents

About the Special Issue Editors

Christos Koulamas, Ph.D. He is currently a Research Director at the Industrial Systems Institute (ISI), ATHENA Research and Innovation Centre in Patras, Greece. He has more than 25 years of experience in R&D, in the areas of real-time distributed embedded systems, industrial networks, wireless sensor networks and their applications in industrial and building automation, and the transportation and energy sectors. He is the author or co-author of more than 60 publications in international journals and conferences and he has participated in numerous European and national research projects with multiple roles. He has served as a member of the board of ERCIM, as Guest Editor for MDPI journals and as PC/TPC or panel member of many IEEE and other conferences and workshops. He is an IEEE Senior Member and a member of the Technical Chamber of Greece.

Mihai Teodor Lazarescu, Assistant Professor at Politecnico di Torino (Italy), received his Ph.D. from same institute in 1998. He was Senior Engineer at Cadence Design Systems, working on high-level synthesis (HLS) of embedded systems, founded several startups working on real-time embedded Linux and WSN for long-term environmental monitoring, and participated in numerous European- and national-founded research projects on topics related to the design of ASIC, EDA for HLS, and WSN hardware, software and development tools. He has authored and co-authored more than 50 scientific publications at international conferences and journals, and several books. As an IEEE member, he served as Guest Editor for several ACM and MDPI journals and as PC Chair for international conferences. His research interests include real-time embedded systems, low-power sensing and data processing for IoT, WSN platforms, high-level hardware/software co-design, and high-level synthesis.

Editorial

Real-Time Embedded Systems: Present and Future

Christos Koulamas [1,*] and Mihai T. Lazarescu [2]

[1]   Industrial Systems Institute/"Athena" Research Center, PSP Bldg, Stadiou Strt, 26504 Patras, Greece
[2]   Dipartimento di Elettronica e Telecomunicazioni Politecnico di Torino Turin, I-10129 Turin, Italy;
       mihai.lazarescu@polito.it
*   Correspondence: koulamas@isi.gr; Tel.: +30-2610-911-597

Received: 10 September 2018; Accepted: 14 September 2018; Published: 18 September 2018

1. Introduction

Real-time and networked embedded systems are important bidirectional bridges between the physical and the information worlds. Embedded intelligence increasingly pervades industry, infrastructure, and public and private spaces, being identified as a society and economy emerging "neural system" that supports both societal changes and economic growth. As cost/performance improves, everyday life connected objects increasingly rely on embedded intelligence in an ever-growing array of application fields, specialized technologies, and engineering disciplines.

While this process gradually builds the Internet of Things (IoT), it exposes a series of specific non-trivial timing and other extra-functional requirements and system properties, which are less common in other computing areas. For instance, most embedded systems are cost-sensitive and with real-time constraints, optimized for power and specific tasks, built around a wide array of processors, often resource-constrained, which need to operate under extreme environmental conditions, and where reliability and security can have severe implications.

The area is quite wide, with diverse computer science and engineering fields and practices involved, and the state of the art is mostly captured today in the cyber-physical systems [1] and IoT [2] evolution contexts, addressing design methods and tools [3], operating systems and resource management [4], real-time wireless networking [5], as well as safety and security [6] aspects, either horizontally or vertically along specific application domains.

2. The Present Issue

The ten articles in this special issue propose solutions to specific open problems of cyber-physical and real-time embedded systems applicable to both traditional application domains, such as industrial automation and control, energy management, automotive, aerospace and defense systems, as well as emerging domains, such as medical devices, household appliances, mobile multimedia, gaming, and entertainment systems. Specifically, they address important topics related to efficient embedded digital signal processing (DSP), security and safety, scheduling, and support for smart electric grid optimizations.

Efficient digital signal processing is an important enabler for advanced embedded applications in many domains. Of the five articles in this special issue that address efficient embedded DSP and applications, the work in Reference [7] studies feasibility options and evaluates the performance of fully embedded algorithms for real-time ventricular fibrillation detectors which can send timely alerts without requiring external processing, for applications in pervasive health monitoring. Health monitoring applications are also the focus of Reference [8] that proposes an efficient embedded hardware accelerator for long-term bio-signal monitoring and compression, which makes it suitable for various Internet of Things (IoT) applications. Hardware-assisted efficient embedded DSPs are essential in other application domains, too. An FPGA implementation of a multi-band real-time speech enhancement system is shown in Reference [9], which includes specific architectural optimizations for speed and energy consumption. An application-specific integrated circuit (ASIC) implementation of real-time and accuracy optimizations for arctangent calculation based on a coordinate rotation digital computer (CORDIC) is shown in Reference [10] (a trigonometric function that is essential for many embedded DSP calculations), while an automated folding scheme for efficient FPGA implementation of fast Fourier transform (FFT), an algorithm widely used in many embedded application domains, is proposed and experimentally evaluated in Reference [11].

As embedded systems pervade most human activities, from households to industry, safety and security architectural and operational aspects become very important. A survey of microarchitectural attacks of embedded systems is presented in Reference [12], which emphasizes potentially very harmful hardware vulnerabilities that usually receive far less attention than the software ones. Middleware often encompasses several embedded and infrastructure systems, which may also be of different types, hence its reliability is highly relevant for the overall security of the applications. In Reference [13], the authors present a formal specification for middleware correctness and a method to thoroughly verify it at runtime. Efficient virtualization for real-time embedded systems is presented in Reference [14], with emphasis on security, safety, functionality and flexible adaptation to most embedded operating systems.

Scheduling for energy efficiency and effective real-time response of embedded systems is addressed in Reference [15], which propose static scheduling methods based on mixed integer linear programming and heuristics, for both periodic and non-periodic tasks.

Finally, as smart electric grids rely (among other requirements) on effective consumer cooperation and coordination to be able to optimize energy production and distribution, the work in Reference [16] proposes several cloud-enabled embedded solutions and provides experimental test and validation results.

3. Future

While embedded systems are not novel, they recently accelerated their permeation in most human activities, which increasingly rely on their real-time capability for, e.g., sensing, processing, communication, actuation, and composability.

Their growing importance for the operation and decision-making capability of complex systems raise significant issues regarding many other aspects of their operation. Security, safety, and reliability may directly impact on physical and data safety as embedded systems are increasingly deployed to measure and control human environments, gaining access, and processing sensitive data. Embedded systems control, validation, self-testing, and observability of their programmed or acquired operation are significant concerns for their acceptance in critical infrastructures or operations, while cost, energy, and maintenance requirements become increasingly important to economically sustain the high number of embedded systems expected for IoT applications. Edge processing, on embedded systems, gains importance over data transfer for remote processing to provide a faster real-time response, reduce dependency on data connections, improve scalability, and increase security by reducing the attack surface.

Recent hardware and software [17] advances aimed at efficiently supporting artificial intelligence (IA) applications directly on resource- and energy-constrained embedded systems for a wide class of applications, compound the importance to find effective solutions to above concerns.

Author Contributions: C.K. and M.T.L. worked together during the whole editorial process of the special issue, "Real-Time Embedded Systems", published in the MDPI journal Electronics. M.T.L. drafted this editorial summary. C.K. and M.T.L. reviewed, edited and finalized the manuscript.

Acknowledgments: We thank all authors, who submitted excellent research work to this special issue. We are grateful to all reviewers who contributed evaluations of scientific merits and quality of the manuscripts and provided countless valuable suggestions to improve their quality and the overall value for the scientific community. Our special thanks go to the editorial board of MDPI Electronics journal for the opportunity to guest edit this special issue, and to the Electronics Editorial Office staff for the hard and precise work to keep a rigorous peer-review schedule and timely publication.

Conflicts of Interest: The authors declare no conflict of interest.

References

1. Khaitan, S.K.; McCalley, J.D. Design Techniques and Applications of Cyberphysical Systems: A Survey. *IEEE Syst. J.* **2015**, *9*, 350–365. [CrossRef]
2. Samie, F.; Bauer, L.; Henkel, J. IoT technologies for embedded computing: A survey. In Proceedings of the Eleventh IEEE/ACM/IFIP International Conference on Hardware/Software Codesign and System Synthesis (CODES '16), New York, NY, USA, 1–7 October 2016.
3. Jensen, J.C.; Chang, D.H.; Lee, E.A. A model-based design methodology for cyber-physical systems. In Proceedings of the 2011 7th International Wireless Communications and Mobile Computing Conference, Istanbul, Turkey, 4–8 July 2011; pp. 1666–1671.
4. Musaddiq, A.; Zikria, Y.B.; Hahm, O.; Yu, H.; Bashir, A.K.; Kim, S.W. A Survey on Resource Management in IoT Operating Systems. *IEEE Access* **2018**, *6*, 8459–8482. [CrossRef]
5. Lu, C.; Saifullah, A.; Li, B.; Sha, M.; Gonzalez, H.; Gunatilaka, D.; Wu, C.; Nie, L.; Chen, Y. Real-Time Wireless Sensor-Actuator Networks for Industrial Cyber-Physical Systems. *Proc. IEEE* **2016**, *104*, 1013–1024. [CrossRef]
6. Wolf, M.; Serpanos, D. Safety and Security in Cyber-Physical Systems and Internet-of-Things Systems. *Proc. IEEE* **2018**, *106*, 9–20. [CrossRef]
7. Kwon, S.; Kim, J.; Chu, C. Real-Time Ventricular Fibrillation Detection Using an Embedded Microcontroller in a Pervasive Environment. *Electronics* **2018**, *7*, 88. [CrossRef]
8. Antonopoulos, C.; Voros, N. A Data Compression Hardware Accelerator Enabling Long-Term Biosignal Monitoring Based on Ultra-Low Power IoT Platforms. *Electronics* **2017**, *6*, 54. [CrossRef]
9. Bahoura, M. Pipelined Architecture of Multi-Band Spectral Subtraction Algorithm for Speech Enhancement. *Electronics* **2017**, *6*, 73. [CrossRef]
10. Pilato, L.; Fanucci, L.; Saponara, S. Real-Time and High-Accuracy Arctangent Computation Using CORDIC and Fast Magnitude Estimation. *Electronics* **2017**, *6*, 22. [CrossRef]
11. Minotta, F.; Jimenez, M.; Rodriguez, D. Automated Scalable Address Generation Patterns for 2-Dimensional Folding Schemes in Radix-2 FFT Implementations. *Electronics* **2018**, *7*, 33. [CrossRef]
12. Fournaris, A.; Pocero Fraile, L.; Koufopavlou, O. Exploiting Hardware Vulnerabilities to Attack Embedded System Devices: A Survey of Potent Microarchitectural Attacks. *Electronics* **2017**, *6*, 52. [CrossRef]
13. Khan, M.; Serpanos, D.; Shrobe, H. A Formally Reliable Cognitive Middleware for the Security of Industrial Control Systems. *Electronics* **2017**, *6*, 58. [CrossRef]
14. Martins, J.; Alves, J.; Cabral, J.; Tavares, A.; Pinto, S. µRTZVisor: A Secure and Safe Real-Time Hypervisor. *Electronics* **2017**, *6*, 93. [CrossRef]
15. Jiang, X.; Huang, K.; Zhang, X.; Yan, R.; Wang, K.; Xiong, D.; Yan, X. Energy-Efficient Scheduling of Periodic Applications on Safety-Critical Time-Triggered Multiprocessor Systems. *Electronics* **2018**, *7*, 98. [CrossRef]
16. Brusco, G.; Burgio, A.; Menniti, D.; Pinnarelli, A.; Sorrentino, N.; Scarcello, L. An Energy Box in a Cloud-Based Architecture for Autonomous Demand Response of Prosumers and Prosumages. *Electronics* **2017**, *6*, 98. [CrossRef]
17. Tang, J.; Sun, D.; Liu, S.; Gaudiot, J.L. Enabling Deep Learning on IoT Devices. *IEEE Comput.* **2017**, *50*, 92–96. [CrossRef]

Article

Real-Time Ventricular Fibrillation Detection Using an Embedded Microcontroller in a Pervasive Environment

Sundeok Kwon [1], Jungyoon Kim [2,*,†] and Chao-Hsien Chu [3]

[1] College of Engineering, Youngsan University, 288, Junam-ro, Yangsan-si 50510, Gyeongsangnam-do, Korea;
    winder2000@naver.com
[2] The Tilbury Research Group, College of Engineering, University of Michigan, Ann Arbor, MI 48109, USA
[3] College of Information Sciences and Technology, The Pennsylvania State University, E327 Westgate Building,
    University Park, PA 16802, USA; chu@ist.psu.edu
*   Correspondence: bassjykim@gmail.com; Tel.: +82-10-5093-6786
†   Current Address: Department of Computer Science, Kent State University, 241 Mathematics and Computer
    Science Building, Kent, OH 44242-0001, USA.

Received: 1 May 2018; Accepted: 30 May 2018; Published: 3 June 2018

Abstract: Many healthcare problems are life threatening and need real-time detection to improve patient safety. Heart attack or ventricular fibrillation (VF) is a common problem worldwide. Most previous research on VF detection has used ECG devices to capture data and sent to other higher performance units for processing and has relied on domain experts and/or sophisticated algorithms for detection. In this case, it delayed the response time and consumed much more energy of the ECG module. In this study, we propose a prototype that an embedded microcontroller where an ECG sensor is used to capture, filter and process data, run VF detection algorithms, and only transmit the detected event to the smartphone for alert and call for services. We discuss how to adapt a common filtering and scale process and five light-weighted algorithms from open literature to realize the idea. We also develop an integrated prototype, which emulates the VF process from existing data sets, to evaluate the detection capability of the framework and algorithms. Our results show that (1) TD outperforms the other four algorithms considered with sensitivity reaching 96.56% and specificity reaching 81.53% in the MIT-BIH dataset. Our evaluations confirm that with some adaptation the conventional filtering process and detection algorithms can be efficiently deployed in a microcontroller with good detection accuracy while saving battery power, shortening response time, and conserving the network bandwidth.

Keywords: real-time detection; wearable ECG device; energy consumption; ventricular fibrillation; VF detection algorithms

---

1. Introduction

Cardiovascular disease (CVD) is life threatening and there is a need for real-time detection to improve patient safety. CVD, caused by the malfunction of heart and blood vessels, is a common problem worldwide. The World Health Organization [1] reported that 17.7 million people died from CVD in 2015, accounting for about 31% of all deaths globally. According to the WHO, if there is no attempt to solve this problem, the rate of death from CVD will increase continuously. Among the various forms of heart disease, ventricular fibrillation (VF) is one of the most severe and dangerous abnormal heart rhythms. Indeed, VF can lead to death within a few minutes or a few days. The survival rate for a person who has a VF outside the hospital ranges between 7% and 70%, depending on the timing of first-aids [2].

Over the past decades, two streams of research have emerged in real-time VF detection: (1) designing wearable devices that can be used to continuously and reliably monitor health conditions [3,4] and (2) developing efficient algorithms that can correctly detect VF abnormality, especially in real-time [5]. However, with limited resources available in microcontrollers and smartphones, several research questions remain unresolved.

First, most of these wearable monitoring devices have no or only have simple detection functions and use wireless transmission to send data to a receiving unit [5]. They lack capability for real-time detection. In addition, the wireless transmission operations consume lots of battery power in comparison to other operations in the microcontroller, and use more network bandwidth [6]. Thus, how to increase diagnosis intelligence and reduce wireless data transmission to increase the battery life of the monitoring devices without replacing or recharging batteries within reasonable time duration is an important issue.

Secondly, previous studies on VF detection have been predominantly focused on evaluating the relative performance of the proposed algorithms in a centrally located computer or a smartphone. Meanwhile, most VF detection algorithms in previous studies used the "filtering.m" filtering process (available online [7]), which consists four successive steps: (1) mean value subtraction, (2) 5th order moving average filter, (3) drift suppression using high pass filter (1 Hz cut-off frequency), and (4) drift high frequency suppression using low pass Butterworth filter (30 Hz cut-off frequency); however, most studies assumed that the process to remove various noises was a preliminary step. That means, this process was not included as part of the evaluation process. Thus, misreporting the detection accuracy.

Thirdly, with the rapid development of mobile devices and microchip sensors, and their increasing usage in e-health, there is also an emerging need to examine how to reduce unavoidable noises in real time [8] in a pervasive environment, especially in a microcontroller.

Accordingly, the following issues still need to be examined: How should we include or deploy these filtering processes and detection algorithms in mobile devices or microcontrollers to reduce unavoidable noises? How effective are these filtering processes? What is the potential power consumption concern? How well do these algorithms perform? Moreover, how can we properly determine the system parameters of the algorithms such as window segmentation and threshold value?

Traditionally, a centrally located machine, cloud or smartphone is used to receive and process the raw ECG data from wearable sensing devices and a human expert or algorithms are then called in to detect VF abnormality. These approaches still need to rely on a microcontroller inside the embedded module to sample the raw digital data produced by the analog-to-digital converter (ADC) and then transmit them to higher performance machines for processing. The data transmission process consume a significant amount of battery power, delays the response time, and exposes the possibility of data leakage. In this study, we propose a methodology that uses an embedded microcontroller as a fog node to filter noise, extract signals and detect VF and only send the detected events to smartphones to alert and request healthcare services. Once an abnormality is detected, the complete heart signal can then be captured and transmitted to a server or cloud for in-depth analysis and treatment. The proposed methodology would shorten the incident response time and significantly reduce the battery consumption of the wearable ECG device, as most of the device's power consumption stems from the abundance of data transmitted via wireless communications [9]. Table 1 summarize the characteristics of these approaches.

As a benchmark, we adapt five light-weighted algorithms from open literature and use the complete set of MIT and Boston's Beth Israel Hospital arrhythmia database (MIT-BIH) and Creighton University (CU) Ventricular Tachyarrhythmia databases [10] for tests. Performance results are measured based on common quality metrics such as sensitivity, selectivity, positive predictivity, accuracy, computational time, power consumption and receiver operating characteristic (ROC) curve.

Table 1. Summary of alternative VF detection approaches.

Tasks/Performance	Traditional Approach	Smartphone-Based	Proposed Approach
Data Capture	EM	EM	EM
Data Transmitted	Raw data	Raw data	Detected events
Transmission Method	SP	SP	SP
Data Filtering and Scale	CCS	SP	EM
Data Analysis	CCS	SP	EM
Event Detection	CCS	SP	EM
Service Request	SP	SP	SP
Response time (delay)	Long	Some	Little
Energy Consumption	High	High	Low
Accuracy	Same	Same	Same
Network Traffic	High	Some	Low

2. Related Studies

There have been many studies focused on evaluating VF detection algorithms for automated external defibrillators (AEDs) [11–16] or mobile phone [17–19]. In addition, there are several studies focusing on designing wearable ECG devices that can transmit data efficiently and securely [5,6,9,20,21].

The most recent and seminal work that compares different algorithms are the studies by Amann et al. [12,13,22], Ismail et al. [14], Abu et al. [15] and Arafat et al. [16]. Amann et al. [12] compared five well-known standard QRS complex and five new VF detection algorithms with a large amount of data that has been annotated by qualified cardiologists. They choose the three annotated databases without pre-selection of certain ECG episodes for tests. Within the results, TCI [23], VFF [24], and TOMP [25] are particularly noteworthy.

In a follow-up study, Amann et al. [22] developed a VF detection algorithm, called Time Delay (TD), based on phase space reconstruction and then evaluated it against four extant algorithms, TCI, VFF, spectral algorithm (SPEC), and complex measure (CPLX), all of which were reported in their previous study [12]. TD, which counts the number of boxes based on the phase-space plots of random signals, can also be easily implemented in a microcontroller [26]. Ismail et al. [14] compared five different algorithms including CPLX, MEA, TCI, TD, and VFF and explored the impact of combining two algorithms. They concluded that combining two algorithms might improve the performance. Abu et al. [15] developed a sequential detection algorithm using empirical mode decomposed (EMD) analysis that showed improved performance over six algorithms including TD and TCI. However, EMD requires relatively high computational complexity comparing with others. Arafat et al. [16] developed a time domain algorithm, called TCSC, adapted from TCI and compared its performance with six algorithms including CPLX, TCI, and TD. They concluded that TCSC performed better than TCI based on positive threshold.

The literature has shown mixed results among detection methods from different studies. This variation could be due to the use of different threshold values for comparison and experimental settings; unfortunately, most studies did not report the threshold parameters used in their studies. However, one clear thing is that TCI, TCSC, TD, VFF, and TOMP are among the popular methods used for benchmark testing. In addition, to the best of our knowledge, there is no study attempting to perform real-time VF detection on microcontrollers, despite that doing so can reduce battery consumption and shorten response time when life-threatening emergencies occur. That is the focus of this study.

In terms of real-time health monitoring, Flores-Mangas and Oliver [6] examined the progress of previous work, revealed system requirements and proposed a framework for real-time monitoring and analyzing physiological signals. They have also used sleep apnea as an example which demonstrated, experimented and evaluated the prototype performance. The paper did not provide details on key information processing tasks such as data filtering and segmentation. Dagtas et al. [9]

presented a 3-tiered wireless architecture and prototype using Zigbee for heart failure monitoring and detection. The paper provided detailed description, noise filtering, basic QRS detection and secured data communication setting. The ideas and implementation are similar to our proposed methodology—using microcontroller (MCU) to capture, process, and detect abnormality and then send detailed signal after detection. However, they only used a simple algorithm to detect QRS (not VF). Meanwhile, no detailed evaluation result has been provided. Pantelopoulos and Bourbakis [5] conducted a state-of-the-art survey on wearable sensor-based systems for health monitoring. They reviewed the current systems developed and identified the technological shortcoming and challenge issues, which can provide directions for future research improvements. Choudhary et al. [20] developed a real-time wireless system based on Zigbee technology for remote monitoring of patients' heart rate and oxygen saturation in blood. The proposed methodology used MCU to store, convert and send signals periodically back to PC for further analysis by users. There is no filtering and detection capability in the MCU unit. Yadav et al. [21] designed a mobile health monitoring system using heart beat and body temperature for potential heart disease detection. The study provided fundamental system framework and illustrated the role of GPS in location detection but it only used simple threshold rule for heart rate and temperature, which cannot accurately detect heart attack and thus the proposed system is not much useful for practical real world situation. Clearly, the development of wearable systems for real-time health monitoring is evolving; however, a well-designed methodology that considers the integration of hardware, software and services for usability, real-time data processing and detection for effective and fast response, lower power consumption, and secure data communications is still expected. In this paper, we studied the integration of hardware and software, real-time data processing and detection for effective and lower power consumption.

3. Methodology

This paper proposes a methodology with three alternative strategies to increase the responsiveness and accuracy of detection and reduce power consumption:

1. Adopt an interrupt service routine (ISR) to reduce power consumption.
2. Use the microcontroller of the embedded ECG module to filter, scale and process the data, and detect VF patterns and only transmit the event via a smartphone to shorten response time and reduce power consumption.
3. Select and adapt VF detection algorithms that have accurate and efficient workloads for the microcontroller used.

3.1. The Proposed VF Detection Methodology

In this study, we propose a methodology that uses the embedded microcontroller as a fog node to filter noise, extract signals and detect VF and only send the detected events to a smartphone to alert and request healthcare services. Once an abnormality is detected, the complete heart signal can then be captured and transmitted to a server or cloud for in-depth analysis and treatment. The proposed methodology would shorten the incident response time and significantly reduce the battery consumption of the wearable ECG device, as most of the device's power consumption stems from the abundance of data transmitted via wireless communications [3].

Figure 1 illustrates the implemented process. The process starts with a timer interrupt for capturing ECG, filtering, and scaling at every sampling point. The captured data is stored in the window storage of the memory. Once the window size, Ws, is filled with the captured data, VF detection algorithm is applied to check whether VF event is occurred or not, and the result is transmitted to the smartphone or main computer through wireless communications.

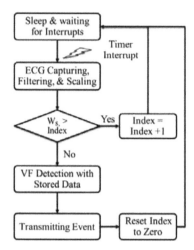

Figure 1. Flow chart of the proposed VF detection methodology.

The power consumption of related operations—signal acquisition, VF detection and data transmission is shown in Figure 2. As can be seen, transmitting operation takes five times more battery power voltage than signal acquisition and detections if we assume the operating duration is the same. In Figure 2, the transmission operation sends only the result event, which is only one value so that it takes short duration relative to acquisition. If the system transmits all the captured data, the transmission consumes much more power than other operations. With our proposed implementation, instead of transmitting all data captured to smartphone, we process VF detection at microcontroller and only transmit the detected events; thus, we can save the transmitting power as in a way proportional to the number of capturing.

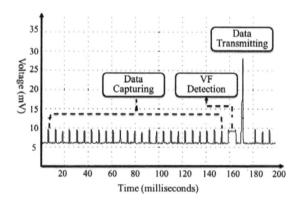

Figure 2. Voltage differences in the proposed VF detection methodology.

3.2. The Filtering Process

Many earlier studies in VF detection used a filtering function, called *filtering.m*, implemented in Matlab [8] and conducted offline. The function contains four sub-routines: mean value subtraction, 5th order moving average to remove high frequency noise like interspersions and muscle noise, drift suppression using high pass filter (1 Hz cut-off frequency), and low pass Butterworth filter (30 Hz cut-off frequency), which suppresses the high frequency noises even more.

The conventional *filtering.m* function, however, does have some disadvantages if the process is implemented in real time using an embedded microprocessor that has limited computational resources and battery energy. First, the battery life will be significantly shortened. Secondly, in real time situations, it is hard to calculate the mean value and moving average of the whole ECG signal. Thirdly, according to our pilot tests, the detection performance is not significantly affected by removing the mean value subtraction and moving average filtering. Therefore, in our implementation, we drop the first two subroutines, adapt the high-pass (1 Hz) and low-pass (30 Hz), and introduce the Kalman filters in the testbed.

3.2.1. High and Low Pass Filter for Removing Noise

The cut off frequency of high pass filter (1 Hz) and low pass filter (30 Hz) can typically be calculated as follows:

$$f_c = \frac{1}{2\pi RC} \tag{1}$$

We can then apply R (Resistor) and C (Capacitor) values to the transfer function of the high pass filter as shown:

$$\frac{V_o}{V_i} = \frac{j\omega RC}{1 + j\omega RC} \tag{2}$$

Using the z-transform, the transfer function can be obtained with R and C values. Although there are several ways to transfer the z-transform, we use the Tustin method with the c2d function available in the Matlab [8]. As a result of this process, the transfer function for the high-pass filter can be obtained with coefficients as shown:

$$Y(z) = 0.9043Y(z - 1) + 0.9522U(z) - 0.9522(z - 1), \tag{3}$$

where $Y(z - 1)$ is the previous output value; $U(z)$ is the current input value; and $U(z - 1)$ is the previous input value. We then apply R and C values to the transfer function of the low-pass filter:

$$\frac{V_o}{V_i} = \frac{1}{1 + j\omega RC}. \tag{4}$$

Using the same process as the high-pass filter, the transfer function for the low-pass filter is obtained with coefficients as follows:

$$Y(z) = -0.2021Y(z - 1) + 0.6011U(z) + 0.6011U(z - 1). \tag{5}$$

3.2.2. Kalman Filter for Tracking Baseline

The threshold-based algorithms such as TCI and TCSC need to track a baseline to have detections that are more accurate. A baseline is a reference line that indicates the trend of an ECG signal without QRS which represents the rapid depolarization of the right and left ventricles and other factors. Thus, we assume those factors as noises when we track the baseline. We use the linear Kalman filter [27], which is a method that calculates a posteriori estimate using:

$$\hat{x}_k \hat{x}_k = \hat{x}_{k-1}^- + K\left(z_k + H\hat{x}_k^-\right), \tag{6}$$

where $\hat{x}_k$ is a posteriori estimate value; K is variable weight; z_k is a measured value and H is a system model parameter. Although the format of Kalman filter is similar to one-dimension low-pass filter, it has a priori estimate, measured value, and dynamic weight K, which recalculated every time.

$$K_k = P_k^- H^T \left(HP_k^- H^T + R\right)^{-1}, \tag{7}$$

where P_k is the error of the posteriori estimate value, which has a linear relationship with the estimate error:

$$P_k = P_k^- - K_k H P_k^-. \tag{8}$$

In addition, the following relationship is established between posteriori estimate value and error covariance value about x_k.

$$x_k \sim N(\hat{x}_k, P_k). \tag{9}$$

The process of Kalman filter can be described in two repeated steps:

(1) Prediction. First, a target system is modeled, based upon which, the Kalman filter predicts a priori estimate and error covariance. Where error covariance is a measure of the error of the posteriori estimate value. The performance of the filter depends on how similar the modeled system is to the actual system.

(2) Estimation. The filter then calculates a new posteriori estimate value based on the difference between the measured value and the priori estimate value. In the Kalman filter, noise is an important variable. The status variable, error covariance and system model are expressed as:

$$\hat{x}_k^- = A\hat{x}_{k-1}^-, \tag{10}$$

$$P_{k+1}^- = AP_k A^T, \tag{11}$$

$$x_{k+1} = Ax_k + w_k, \tag{12}$$

$$z_k = Hx_k + v_k. \tag{13}$$

In this paper, we set the parameters, based on our tests using CU database, as: $A = 1, H = 1, Q = 0$ and $R = 4$.

3.3. The Scaling Process

The amplitude of the ECG signal from a human body is normally about 1.3 mV. However, the minute ECG signal is often amplified through the front-end device and active filters to fit the total range of power. The voltage of the microprocessor is recently 3 V and the analog-to-digital converter has resolutions of 8–12 bits. The larger the amplified ECG signal that is achieved, the better the detecting accuracy will be. Thus, we can assume that the raw ECG signal (the range of 1 mV~5 mV) is simply amplified to fit into the maximum limitation. However, there is always a short-term or long-term drift noise. If the baseline noise is not taken into account during the amplification process, a clipping fault in which the signal exceeds the limited range of the ADC can occur.

In VF detection, the amplitude of ECG signal is varied in the cases of sinus and VF signals. Scaling is a process by which ECG signals of a window segment with different peak-to-peak amplitudes are stretched into one uniform amplitude size. If there is no scaling, the parameter values vary from the size of peak-to-peak amplitudes. In real-time detection, the variances are unpredictable, meaning that errors may occur with the wrong thresholds. To obtain accurate thresholds, the periodic changing of the maximum and minimum values of the scale in the segment blocks is required. If the period is too short, the algorithm may lose the QRS complexes. On the other hand, if the period is too long, small VF events can be detected as sinus rhythms. We apply a scaling process for every window segment, which is 5 seconds in our implementation. Figure 3 shows the proposed scaling process.

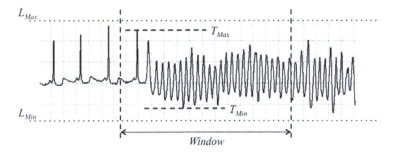

Figure 3. An example of the scaling process.

The scaled value V_t can be calculated as:

$$V_t = S_t V_{Mul} - V_{Add}, \tag{14}$$

$$V_{Mul} = \frac{L_{Max} - L_{Min}}{T_{Max} - T_{Min}}, \tag{15}$$

$$V_{Add} = |T_{Max} - T_{Min}|, \tag{16}$$

where S_t is raw ECG data; V_{Mul} is the value for multiplication; and V_{Add} is the value for addition. The V_{Mul} and V_{Add} values are calculated simply using Equations (15) and (16). L_{Max} is the maximum limitation and L_{Min} is the minimum limitation. Since we use an 8-bit sampling ADC resolution, the L_{Min} is zero and the L_{Max} is 255 (i.e., $2^8 - 1$). If a 10-bit resolution was used, then the L_{Min} is zero and the L_{Max} is 1023 (i.e., $2^{10} - 1$). T_{Max} is the maximum value of signal; and T_{Min} is the minimum value of signal.

3.4. The Revision in Detection Algorithms

Due to the memory limitation of microcontroller, every algorithm in this study uses a 5-s window size to fit in a temporal memory. Based on the window size, we slightly revise the algorithms and determine the threshold of VF and SR. We discuss each algorithm with the revision below.

3.4.1. The TCI algorithm

TCI [23] calculates the threshold value (T), the number of crossing signals (C) and the number of intervals (I) in three consecutive 1-second blocks to determine the VF or non-VF decision based on the number and the position of the crossings. *TCI* estimates the average interval between ECG signal threshold crossings. The mean *TCI* value is calculated as:

$$TCI = \frac{1000}{(N-1) + \frac{t_2}{t_1+t_2} + \frac{t_3}{t_3+t_4}}, \tag{17}$$

where N is the number of impulses in segment S (1 s interval) and $t_1 \sim t_4$ are intervals between crossing points and each end points of the S. If a prescribed threshold value, TCI_0, is greater than the *TCI* value, VF is declared. There is no revision applied to this algorithm.

3.4.2. The TCSC Algorithm

TCSC [16] is an improved method over the TCI algorithm, which removes the drawbacks of TCI such as using 1-second blocks as analysis windows that causes trouble when the heartbeat is fewer than 60 bpm and one threshold (20% of the maximum value) that causes a missing problem with negative peak value of ECG. Three additional extensions TCSC adds over TCI are:

11

- A 3-second block is investigated instead of a 1-second block.
- Both positive and negative thresholds are used instead of only the positive threshold.
- Samples above the thresholds are counted instead of only the pulses.

Our proposed revision: In the original TCSC algorithm, each segment is multiplied by a cosine window $w(t)$ that minimizes the effect from end-to-end of each segment. We skip the multiplication of the cosine window, which only minimize the effects of amplitudes in the beginning and ending 0.25 s of each window segment (see Figure 4), but it needs high computational math functions compared with the performance in a microcontroller. Based on our experiment, the impact on signal detection is minimized but on computational time is significant. For instance, the computational time with cosine window in one window segment (5 s in our case) is 17.84 ms but it only takes 6.8 ms without using the cosine window.

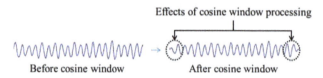

Figure 4. Effects of cosine window process in TCSC.

3.4.3. The TD Algorithm

TD [22] uses a 40 × 40 squares grid of the two-dimensional phase space diagram that refers to different types of plotting using normal data $X(t)$ as the x-axis and $X(t + \tau)$ as the y-axis, with $\tau = 0.5$ s. The values are plotted in the 1600 squares and the resultant parameter is the number of visited boxes. To distinguish QRS complex from VF cases, TD counts the density of visited boxes and compares it with a threshold value. If the calculated density is less than a prescribed threshold density, the ECG signal is considered as a normal QRS complex; otherwise, it will be classified as a VF case. In general, this algorithm performs well in detecting VF. We skip the process of computing a density value as the step is just divided by the total number of all boxes visited using a floating point calculation, which is more computational intensive and redundant. Moreover, we assign the one box into one single bit so that memory usage is minimized.

3.4.4. The VFF Algorithm

VFF [24] uses a narrow band elimination filter in the scope of the mean frequency of the ECG signal for analysis. After the scaling process, Equations (18) and (19) are applied to the ECG data to get the VF filter leakage l:

$$N = \pi \left(\sum_{i=1}^{m}|V_i|\right)\left(\sum_{i=1}^{m}|V_i - V_{i-1}|\right)^{-1} + \frac{1}{2}, \tag{18}$$

$$l = \left(\sum_{i=1}^{m}|V_i + V_{i-N}|\right)\left(\sum_{i=1}^{m}|V_i| + |V_{i-N}|\right)^{-1}. \tag{19}$$

Once the VF occurs, the window segment is shifted by half a period. If the data in a segment is similar to a periodic signal like VF, this algorithm cancels it so that the VF-filter leakage becomes small. The signal peak of QRS complexes affects the threshold of the VFF algorithm. If the signal is higher than the peak of QRS complex from the previous window segment, the threshold is set as 0.406. Otherwise, the threshold is set as 0.625. There is no revision applied to this algorithm.

3.4.5. The TOMP Algorithm

TOMP [25] is based on a real-time detection method of QRS complex. It applied a squaring function and a moving-window integration to detect QRS complexes. Since the sliding integration window with a width of 150 ms is applied, the slope γ (j) of the ECG data $x(j)$ can be calculated as:

$$\gamma(nT) = \frac{1}{8T}(-x(nT - 2T) - 2x(nT - T) + 2x(nT + T) + x(nT + 2T)), \tag{20}$$

where n is the sample number and T is the sampling period of the ECG signal. The sliding integration window moves and sums the absolute values of the difference between current data and the previous data in a width of 150 ms. The short period of sliding window captures the high peak QRS complexes in ECG data. Using this method, even positive or negative QRS complexes can be easily detected by this sliding window. The TOMP algorithm sets two thresholds about the number of QRS complexes, such as $l_0 = 2$ and $l_1 = 32$. If the number of QRS complexes is out of this range ($l_0 > l$ and $l_1 < l$), it is diagnosed as VF. Thus, it is very important to determine the offset value for deciding whether a VF wave is detected or not.

In general, there is no clear symptom that can indicate when the VF event will happen because the ECG signals normally have no pre-noticeable changing point from the SR (normal ECG) to the VF (as shown in Figure 5). In addition, SR and VF obviously have different features that distinguish QRS complexes from VF signal like a sine wave (or a cosine wave).

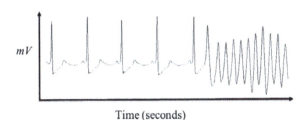

Figure 5. The turning point between a sinus rhythm and the VF event.

Furthermore, QRS complexes vary from signal to signal and are hard to predict, too. For instance, Figure 6 shows different shapes of the QRS complexes in the CU database. In addition, when the VF event occurs, there is no QRS complex in the ECG signal. Therefore, the threshold-based algorithms have difficulty of detecting the various types of QRS complexes.

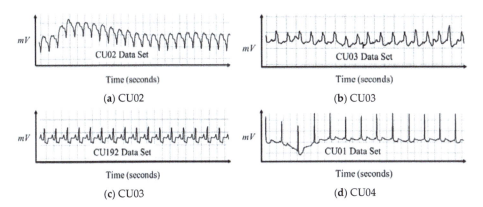

Figure 6. Different types of sinuous rhythms from CU database (62.5 Hz).

3.5. The Development of Testing Prototype

We develop an integrated sensing module as a testbed, based upon which to test the performance and energy consumption of the VF detection methods. Figure 7 shows the conceptual framework of the proposed testing environment.

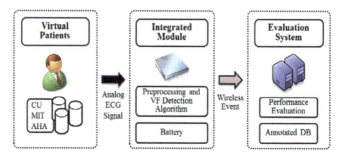

Figure 7. Conceptual framework of the testing environment.

The proposed testing environment, or so-called artificial ECG prototype, contains three modules: (1) virtual patients that house the available testing databases and enable the generation of an analog ECG signal to provide ECG signal with VF events. The artificial ECG can be used to serve as a source to evaluate the VF detection algorithms and the integrated preprocessing filters; (2) the integrated module, which has the VF detection algorithms, hardware and software filtering logic built in. The module receives the ECG signal, removes the noise, and judges whether VF events are occurring. The results from the integrated module are then sent to the evaluation system through wireless communication; and (3) the evaluation system. In the evaluation system, the results of VF detection are compared with previous annotated information, which is then used to calculate the quality parameters.

Figure 8 depicts more specific block diagram of the testing environment. The centralized computer includes the virtual patients and evaluation system. The virtual patients' digital data is converted from analog signals using a digital-to-analog (D/A) converter. Preprocessing in this system consists of three steps: a high pass filter, a low pass filter, and a Kalman filter. The VF detection events are sent to the centralized computer or mobile phone through RF-communication chip, which is TI's CC2500 [28].

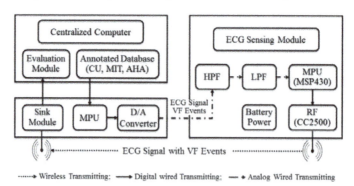

Figure 8. Block diagram of the testing environment.

4. Performance Evaluation

4.1. Experimental Design

We conduct two experiments to verify the feasibility and effectiveness of our proposed development. In experiment 1, we deploy our virtual patients and evaluation system in a laptop computer and evaluate the performance of our algorithms using an 8-second window length, so that our results can be directly compared with those reported in the open literature. This comparison helps to assess the effectiveness of the proposed filtering process and verifies the coding accuracy. In experiment 2, we conduct evaluations using the testing prototype. The same filtering process and algorithms are deployed in the microcontroller unit with minor modifications to meet microcontroller memory requirements (as discussed in Sections 3.2 and 3.4). Signals are extracted every 5 s. Results are then compared with those obtained from experiment 1. This comparison helps to assess the effectiveness of the filtering process and algorithms performed under a microcontroller. It also helps to estimate and justify the potential power saving of our proposed approach.

4.2. The Prototype and Real-Time Evaluation Program

VF detection algorithms need a preprocessing (scaling and filtering) step to improve their detecting performance. We use TI's MSP430-2500 (Texas Instruments, Dallas, TX, USA), which consists of MSP430f2274 microcontroller (Texas Instruments, Dallas, TX, USA) with 32 KB + 256 B of Flash Memory (ROM) and 1 KB of RAM, and CC2500 radio chip (Texas Instruments, Dallas, TX, USA) using 2.4 GHz, to deploy our integrated module. Figure 9 shows the final optimized embedded prototype that we developed. The size of the module is relatively small, which is slightly bigger than a US quarter coin. It can be inserted into a watch type case or into a cloth built in case. The current module has multiple sensors built in, including ECG sensor, temperature sensor (inside the MPU), and accelerometer (i.e., motion detection sensor). The chip antenna and RF transceiver are used for wireless data communication. The MPU contains ADC, memory to store filtering and detection software and captured data, and perform timely analysis and detection. The programming port, which can be removed later, is used to input and update the software programs such as filtering and detection algorithms.

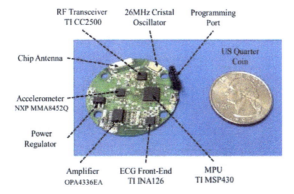

Figure 9. Final optimal prototype of the ECG embedded module.

The real-time evaluation system is coded in C# programming language running in a laptop with Windows 7 operating system. We classify ECG signals into three classes: VF, SR and no VF-SR (Note: There are several periods in the ECG signal marked as "-", which we labeled them as no VF-SR here). Determining the proper threshold parameter is essential but difficult to do. We use a simple discriminant analysis to aid the estimation. According to our tests, most of the algorithms show relatively high reparability between the decisions of SR and VF. The thresholds of each algorithm can be decided at the boundary between the density distributions of SR and VF. Figure 10 shows the probability histograms of the parameter $N_{Time\ delay}$ within the CU database. For example, in this case, the best threshold value for TD is around 145. That is, $N_{Time\ delay} = 145$. Using similar approach, we obtain the threshold values for other algorithms as follows: $N_{TCSC} = 15$, $N_{TOMP} = 10$, and $N_{VF\ filter} = 0.5060$.

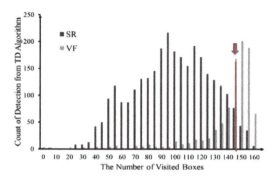

Figure 10. Probability histogram for deciding the parameter of TD within CU database.

4.3. Performance Measures

Four quality performance statistics, sensitivity, specificity, positive predictivity and accuracy, which are commonly employed in automated ECG analysis [7,12–17,29], ROC curve [30] and two other measures related to computation and energy efficiency are also considered to assess performance.

The detecting ability of the algorithms can be compared to the annotations on the database. The comparing process produces four values from raw data: false positives (*FP*), false negatives (*FN*), true positives (*TP*), and true negatives (*TN*). Using these values, various performance indexes can be derived. These measures have been commonly used in evaluating the VF detection research. In order to calculate these measures, four types of values from the systems are collected: false positives (*FP*), false negatives (*FN*), true positives (*TP*), and true negatives (*TN*). Using these values, various performance measures can be derived. The metrics (*Sn, Sp, Pp,* and *Ac*) can then be calculated as follows:

$$Sn = \text{(detected cases of VF)/(all cases of VF)} = TP/(TP + FN), \tag{21}$$

$$Sp = \text{(detected cases of no VF)/(all cases of no VF)} = TN/(TN + FP), \tag{22}$$

$$Pp = \text{(detected cases of VF)/(all cases classified by the algorithm as VF)} = TP/(TP + FP), \tag{23}$$

$$Ac = \text{(all true decisions)/(all decisions)} = (TP + TN)/(TP + FP + TN + FN). \tag{24}$$

In general, a good detection algorithm should produce high sensitivity as well as high specificity. However, sometime this is hard to achieve and needs trade-off due to data imbalance issue commonly found in medical diagnosis problem [30]. The ROC curve, which plots the values of sensitivity and (1–specificity) together in the diagram, is used as a single quality parameter to gauge the quality performance for VF detection.

The computational time, *Ct*, for different algorithms to perform the analysis is recorded in percent rounded to 2 digits relative to the total operation time to assess computational efficiency. In a pervasive sensing monitoring, energy consumption is another important factor for the mobile device. To calculate the average energy consumption for the VF detection algorithms, the integral of the power curve and execution time is needed. We calculate the power value from measured voltage and load resistor. *ET* is the total energy consumption in Joule (J), which is a derived unit of energy. *T* is the total executed time of the database. *Vt* and *It* are the voltage and current values at each time *t*, respectively. For example, if the average power consumption ($P = V \times I$) is 0.3 mW (10 mV $\times$ 0.03 A) with total executed time 300-seconds, E300 is calculated as 90 mJ (0.3 mW $\times$ 300 s).

4.4. The Testing Database

To verify the proposed algorithms, it is very important that the correct annotations suggested by cardiologists can be compared with the decisions between VF and no-VF events derived from the algorithms. Two annotated databases, MIT-BIH (VF) and CU [9], which have been widely used in VF studies [10–13,15,16], are used in this study. The types of rhythms included in these databases are normal sinus rhythm, ventricular tachycardia, and ventricular fibrillation. We review the key characteristics and revise some (mainly sampling frequency (62.5 Hz) and ADC resolution (8 bit)) to perform our evaluation. The key features are summarized in Table 2, where the values in boldface are those we have changed.

Table 2. Summary of performance for VF detection algorithms.

Features	MIT-BIH (VF) Database		CU Database	
	Original	Modified	Original	Modified
Sampling frequency	250 Hz	**62.5 Hz**	250 Hz	**62.5 Hz**
Channel	2	**1**	1	1
ADC Resolution	12 bits	**8 bits**	12 bits	**8 bits**
Number of patients	22	22	35	35
Record length	35 min	35 min	508 s	508 s
Gain (adu/Mv)	200	200	200	200

5. Results and Analysis

5.1. Performance of the Proposed Filtering Method

Table 3a,b summarize the average computational results from experiment 1 using the CU and MIT-BIH (VF) databases, respectively. The boldface values indicate the two best results and the values in italic style or in red color are the two worst results. We have included the corresponding results from three open literatures for comparison. We excluded the most recent literature [14], because they evaluate three databases (CU databae, MIT database, VF database) but used average values for the four quality parameters. If the result is shown as average values of all databases, the accuracy of quality parameters is hard to compare because the results from each database are quite different, for example, as can be seen from Table 3a,b, the results between CU database is quite different from those of MIT-BIH database.

As shown, there is no single algorithm stands out in all four measures either from literature or from our study. However, literature and this study both indicate that VFF algorithm performs better than other algorithms in terms of specificity, positive predictivity, and accuracy; but, it does not do well on the sensitivity measure, which is a major deficiency of VFF. This is because that most data sets, especially the MIT-BIH (VF) data sets, have much more SR signals than VF events. In terms of sensitivity measure, both results show that other algorithms, especially TCI and TCSC, performed much better than VFF.

Overall, in terms of all four quality measures, TCSC and TD performed better than other methods and TOMP performed the worst. In terms of calculation time, early literature did not consider this issue except Amann et al. [11]. Our study indicated that TCSC and TCI are more computationally efficient, followed by VFF, TOMP and TD.

Please note that, in addition to reveal the above similar trend, most of the corresponding values between literature and this study are relatively closer to each other. This assures us that the performance of the proposed filtering process is comparable to the popular *filtering.m* implementation. The results also help to verify the accuracy of our coding process for the algorithms that we adopted. The potential variations in values between literature and this study are due to the differences in several factors (e.g., parameters, sampling frequency, etc.) that may impact the results and unfortunately, most references do not provide specific information about those factors; thus, it is impossible to simulate and verify the exact values from open literature.

Table 3. Summary of performance for VF detection algorithms.

(a) CU Database						
Algorithm	Reference	*Sn* (%)	*Sp* (%)	*Pp* (%)	*Ac* (%)	*Ct* (Second) *
TCI	[11]	71.00	70.50	38.90	70.60	2.1
	[15]	**90.15**	*55.12*	35.70	62.71	-
	[13]	67.89	75.45	-	-	-
	This Study	69.74	62.39	38.31	64.21	**22.40**
VFF	[11]	30.80	**99.50**	**94.50**	85.20	1.9
	[13]	71.41	79.88	-	-	-
	This Study	*36.23*	**99.67**	**97.16**	84.44	42.67
TOMP	[11]	71.30	*48.40*	*26.70*	*53.20*	0.8
	This Study	73.5	54.85	*34.63*	*59.43*	78.23
TD	[21]	70.20	89.30	65.00	85.10	1.7
	[15]	75.35	91.46	70.92	**87.97**	-
	[13]	72.51	88.08	-	-	-
	This Study	69.6	88.26	65.48	83.73	81.79
TCSC	[15]	**79.74**	88.14	65.02	**86.32**	-
	This Study	*63.24*	81.29	51.90	76.92	24.18
(b) MIT-BIH (VF) Database						
Algorithm	References	*Sn* (%)	*Sp* (%)	*Pp* (%)	*Ac* (%)	*Ct* (Second) *
TCI	[11]	74.50	83.90	0.80	83.90	-
	[15]	**100**	*56.82*	*0.37*	*56.89*	-
	This Study	68.39	58.87	38.64	61.48	**58.21**
VFF	[11]	*29.40*	**100.0**	**82.40**	**99.90**	-
	This Study	59.81	96.73	87.06	86.81	110.88
TOMP	[11]	68.50	*40.60*	*0.20*	*40.60*	-
	This Study	88.95	84.97	69.86	86.09	203.28
TD	[21]	74.80	99.20	13.40	99.20	-
	[15]	95.32	99.04	13.83	99.04	-
	This Study	94.01	84.15	70.01	86.94	212.52
TCSC	[15]	**97.48**	**99.33**	18.98	**99.33**	-
	This Study	76.51	72.91	51.77	73.90	**62.83**

* The comparison of *Ct* may not be meaningful because different computers and operating systems are used. Also, previous studies did not include the time for the filtering process used. The boldface values indicate the two best results and the values in *italic* style or in red color are the two worst results.

5.2. Feasibility and Effectiveness

Table 4a,b summarize the average computational results from experiment 2 running on a microcontroller. Where, the boldface values indicate the best results and the italic values (or in red color) indicate their performance is worse than 50%. As shown, most of the values are slightly lower, but not significantly lower, than the corresponding values using 8-s window size. The results give us confidence that both the revised filtering process and the VF detection algorithms were efficiently deployed in a microcontroller with good performance. The slightly degradation in performance can be attributed to the shorter window size used in data extraction.

Table 4. Summary of performance for different VF detection algorithms running in a microcontroller.

(a) CU Database (508 s/each patient data)						
Algorithm	Sn (%)	Sp (%)	Pp (%)	Ac (%)	Ct (%)	$E_{508\text{-}S}$ (mJ)
TCI	**71.63**	61.75	*48.58*	63.65	0.13	109.77
VFF	*38.01*	**99.70**	80.11	84.61	0.24	111.99
TOMP	**77.60**	55.62	45.85	59.63	0.44	115.88
TD	70.88	**88.12**	*72.75*	**84.01**	0.46	116.27
TCSC	62.87	80.87	64.33	77.50	**0.14**	**109.97**
(b) MIT-BIH (VF) Database (35 s/each patient data)						
Algorithm	Sn (%)	Sp (%)	Pp (%)	Ac (%)	Ct (%)	$E_{35\text{-}M}$ (mJ)
TCI	60.66	64.90	38.86	63.76	60.66	**284.97**
VFF	59.94	**96.78**	87.09	86.97	0.24	290.72
TOMP	71.66	95.61	85.70	**89.17**	0.44	300.82
TD	**96.56**	81.53	66.74	85.70	0.46	301.83
TCSC	72.76	72.47	*49.41*	72.55	**0.14**	**285.47**

The boldface values indicate the two best results and the values in *italic* style or in red color are the two worst results.

Similar to the results of experiment 1, the results indicate that VFF is performed the best in terms of specificity, positive predictivity and accuracy for both databases, VFF did not perform well in sensitivity measure; therefore, VFF cannot be the best method for VF detection. On the other hand, TD algorithm is performed quite well in all quality measures (but not necessarily the best for all measures). We need to have a more objective way of choosing proper method, which is where the receiver operating characteristic (ROC) curve was called in. ROC curve takes into consideration of both sensitivity and specificity.

Figure 11 shows the results of ROC curve. In general, the closer the ROC curves to the left-upper point, the better the performance. As shown, the TD algorithm shows the best performance, followed by the TOMP and TCSC algorithms. Both VFF and TCI show some degree of low sensitivity, which indicates that they have difficulty of detecting QRS signals. Although TD has the best performance according to ROC curve, it is still not that close to the left-upper point, which indicates that there are potentials for further improvement.

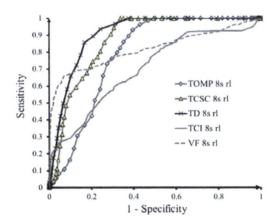

Figure 11. ROC curves for five VF detection algorithms (rl: real-time filter).

5.3. Efficiency and Power Saving

To assess computational time and energy consumption, we compute the relative value (in %) as relative to the lowest value:

$$\text{Relative Value} = \text{Its Value}/\text{Lowest Value}, \tag{25}$$

The lower value means more computational time or a lower energy consumption rate. As shown in Table 3, we can conclude that TCI is the most efficient method for both databases, followed by TCSC. TOMP and TD often took a longer time to obtain results. For easy comparison, we further plot the computational time results and energy consumption in graphs. Figures 12 and 13 show the computational time and energy consumption plots respectively for different VF detection algorithms running in a microcontroller. As shown, TCI consumes less time and battery power than other methods in both databases, followed by TCSC and VF. For both databases, the TD algorithm does consume the most time and energy as shown in the figures. This is because TD algorithm uses 1600 boxes that need to store information in the memory with loop functions for detecting a VF event.

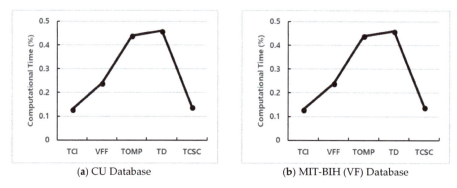

Figure 12. Relative calculation time.

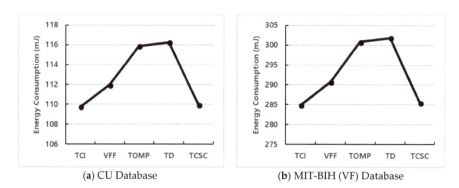

Figure 13. Relative energy consumption.

6. Discussion

Designing wearable systems for real-time health monitoring have to consider many technical, wearable, and medical requirements. For instance, power consumption, mobility, reliability, real-time response, multi-function integration, detection accuracy, wearability, usability, data security and privacy, and clinical validation are important issues to be considered [5,6,9,20]. In addition, we should also consider the resource constraints and computational capability of microcontroller as we use microcontroller for real-time data capturing, filtering, and VF detection. We discuss some specific issues related to microcontroller in more details below.

The first issue, which relates to signal extraction, is the window size to segment the data. In this study, we choose the 5 s window segmentation for data extraction on a microcontroller. If we use more than 5 s as window for segmentation, the memory of the microcontroller is not enough to support the window size. Using 5 s window segment, processing time and detection results are faster than 8 s. However, power consumption is increased because the VF detection algorithm is applied more frequently. Extracting signals every 5 s means that about 312 sampling ECG data (62.5 Hz sampling frequency) are saved in the memory for extracting the features of VF detection. If we use high sampling frequency (250 Hz), the power and processing time for extraction are increased about four times.

The second issue that is critical to quality performance is the determination of prescribed threshold values for different detection algorithms. TCI uses basically a 3 s segment to detect VF events. TOMP, TD, and TCSC count the number of detected events in the window. If accurate thresholds for the changed window size are determined, results similar to the 8 s window can be achieved in the 5 s window size. Although it is possible to use a microcontroller with larger memory to increase the length of window segmentation and thus, it might (or might not) slightly increase the detection accuracy, it will shorten the battery life significantly and result in a longer time to process and detect a VF abnormality. Since our goal is to detect VF in real time, we need to consider the trade-off among duration to detect, battery life, and accuracy.

Thirdly, we face the dilemma of selecting the best algorithm. Theoretically, the best algorithm is the one that provides 100% (or near) results on all quality measures or at least for both sensitivity and specificity measures and yet computationally efficient and less power consumption. However, due to data imbalance problem (that is, most databases consist of a majority of SR signal and very small portion of VF events), it is hard to find a method performed well on all measures. For instance, VFF has good performance in Sp, Pp, and Ac, but it has the worst performance in Sn. According to our tests, TD has good performance in all four quality measures, but it is also more computational intensive and consumes more energy. Considering the need of accurate and robust detection for VF, we would recommend TD for adoption consideration.

Fourthly, It is very important that the trade-off between specificity and sensitivity in safety critical systems such as VF detection. High sensitivity is important issue because missed events will likely lead to death whereas specificity might be also very important if the algorithm is to control the automatic external defibrillator (AED). High specificity is able to prevent the unnecessary shocks from the AED. Thus, selecting the right threshold to more specificity is important in VF detection.

7. Conclusions

This paper proposes an energy efficient real time VF detection methodology with the integration of filtering processes, data extraction and VF detection algorithms in an embedded microcontroller. The results show that real time VF detection using an embedded microcontroller is about five times more efficient than the continuous transmission of ECG signals via wireless communications. Among the selected VF detection algorithms, overall, the TD and TCSC have a relatively high performance considering all the quality parameters. The proposed integrated hardware and software module (ECG device plus microcontroller) can be implemented as an efficient and practical detection system to correctly detect VF abnormality, especially on a real time basis.

To the best of our knowledge, this is one of the first studies that attempts to detect VF in a microcontroller with sophisticated detection algorithms. The main advantages are to save battery power consumption and shorten response time when an emergency occurs. We suggest a revision in the filtering process and some detection algorithms and use a 5-s window segment to ensure that we meet memory requirements. For future studies, first, the impact of different filtering processes needs to be explored in order to select the best filtering method. Secondly, how to reduce window length in a microcontroller needs to be explored. Thirdly, in order to achieve more efficient energy consumption, the hardware implementation of the filtering processes on-chip may need to be considered. Fourthly, to avoid the misdetection of heart activity, processing with multi-sensors such as accelerometers and body temperature sensors need to be examined. Finally, the proposed methodology (The embedded module and integrated testing prototype) can be extended to other applications related to diverse physiological data analysis.

Author Contributions: J.K. and C.-H.C. designed the algorithm and the experiments. S.K. and J.K. implemented the algorithm, analyzed the data and wrote the article. C.-H.C. reviewed the paper and contributed to analysis of the data.

Acknowledgments: This work was supported by a 2016 research grant from Youngsan University, Republic of Korea.

Conflicts of Interest: The authors declare no conflict of interest.

References

1. World Health Organization. Available online: http://www.who.int/mediacentre/factsheets/fs317/en/index.html (accessed on 1 May 2018).
2. Weaver, W.D.; Cobb, L.A.; Hallstrom, A.P.; Copass, M.K.; Ray, R.; Emery, M.; Fahrenbruch, C. Considerations for Improving Survival from Out-of-Hospital Cardiac Arrest. *Ann. Emerg. Med.* **1986**, *15*, 1181–1186. [CrossRef]
3. Leijdekkers, P.; Gay, V. A Self-test to Detect a Heart Attack Using a Mobile Phone and Wearable Sensors. In Proceedings of the 2008 21st IEEE International Symposium on Computer-Based Medical Systems, University of Jyväskylä, Jyvaskyla, Finland, 17–19 June 2008; pp. 93–98.
4. Pantelopoulos, A.; Bourbakis, N.G. A Survey on Wearable Sensor-Based Systems for Health Monitoring and Prognosis. *IEEE Trans. Syst. Man Cybern. C Appl. Rev.* **2010**, *40*, 1–12. [CrossRef]
5. Flores-Mangas, F.; Oliver, N. Healthgear: A Real-Time Wearable System for Monitoring and Analyzing Physiological Signals. In Proceedings of the International Workshop on Wearable and Implantable Body Sensor Networks (BSN'06), Cambridge, MA, USA, 3–5 April 2006.
6. Kim, J.; Chu, C.H. Analysis and Modeling of Selected Energy Consumption Factors for Embedded ECG Devices. *IEEE Sens. J.* **2015**, *16*, 1795–1805. [CrossRef]
7. Filtering.m. Available online: https://homepages.fhv.at/ku/karl/VF/filtering.m (accessed on 1 May 2018).
8. Dağtaş, S.; Pekhteryev, G.; Şahinoğlu, Z.; Çam, H.; Challa, N. Real-Time and Secure Wireless Health Monitoring. *Int. J. Telemed. Appl.* **2008**, *2008*, 1–10. [CrossRef] [PubMed]
9. Khan, J.M.; Katz, R.H.; Pister, K.S.J. Emerging Challenges: Mobile Networking for Smart Dust. *J. Commun. Netw.* **2000**, *2*, 188–196. [CrossRef]
10. Goldberger, A.L.; Amaral, L.A.; Glass, L.; Hausdorff, J.M.; Ivanov, P.C.; Mark, R.G.; Mietus, J.E.; Moody, G.B.; Peng, C.K.; Stanley, H.E. PhysioBank, PhysioToolkit, and PhysioNet: Components of a New Research Resource for Complex Physiologic Signals. *Circulation* **2000**, *101*, E215–E220. [CrossRef] [PubMed]
11. Jekova, I.; Krasteva, V. Real Time Detection of Ventricular Fibrillation and Tachycardia. *Physiol. Meas.* **2004**, *25*, 1167–1178. [CrossRef] [PubMed]
12. Amann, A.; Tratnig, R.; Unterkofler, K. Reliability of Old and New Ventricular Fibrillation Detection Algorithms for Automated External Defibrillators. *BioMed. Eng. OnLine* **2005**, *4*, 60. [CrossRef] [PubMed]
13. Amann, A.; Tratniq, R.; Unterkofler, K. A New Ventricular Fibrillation Detection Algorithm for Automated External Defibrillators. In Proceedings of the Computers in Cardiology, Lyon, France, 25–28 September 2005; pp. 559–562.
14. Ismail, A.H.; Fries, M.; Rossaint, R.; Leonhardt, S. Validating the Reliability of Five Ventricular Fibrillation Detecting Algorithms. In Proceedings of the 4th European Conference of the International Federation for Medical and Biological Engineering, Antwerp, Belgium, 23–27 November 2008; pp. 26–29.
15. Abu, E.M.A.; Lee, S.Y.; Hasan, M.K. Sequential Algorithm for Life Threatening Cardiac Pathologies Detection based on Mean Signal Strength and EMD Functions. *BioMed. Eng. OnLine* **2010**, *9*, 43.
16. Arafat, M.A.; Chowdhury, A.W.; Hasan, M.K. A Simple Time Domain Algorithm for the Detection of Ventricular Fibrillation in Electrocardiogram. *Signal Image Video Process.* **2011**, *5*, 1–10. [CrossRef]

17. Fokhenrood, S.; Leijdekkers, P.; Gay, V. Ventricular Tachycardia/Fibrillation Detection Algorithm for 24/7 Personal Wireless Heart Monitoring. In Proceedings of the 5th International Conference on Smart Homes and Health Telematics, Nara, Japan, 21–23 June 2007; pp. 110–120.

18. Zhang, Z.X.; Tian, X.W.; Lim, J.S. Real-Time Algorithms for a Mobile Cardiac Monitoring System to Detect Life-Threatening Arrhythmias. In Proceedings of the 2nd International Conference on Computer and Automation Engineering, Singapore, 26–28 February 2010; pp. 232–236.

19. Rospierski, M.; Segura, M.; Guzzo, M.; Zavalla, E.; Sisterna, C.; Laciar, E. Ventricular Fibrillation Detection Algorithm Implemented in a Cell-Phone Platform. In Proceedings of the Congreso Argentino de Sistemas Embebidos 2011, Buenos Aires, Argentina, 2–4 March 2011; pp. 168–173.

20. Choudhary, D.; Kumar, R.; Gupta, N. Real-Time Health Monitoring System on Wireless Sensor Network. *Int. J. Adv. Innov. Thoughts Ideas* **2012**, *1*, 37–43.

21. Yadav, D.; Agrawal, M.; Bhatnagar, U.; Gupta, S. Real Time Health Monitoring Using GPRS Technology. *Int. J. Recent Innov. Trends Comput. Commun.* **2013**, *1*, 368–372.

22. Amann, A.; Tratniq, R.; Unterkofler, K. Detecting Ventricular Fibrillation by Time-Delay Methods. *IEEE Trans. Biomed. Eng.* **2007**, *54*, 174–177. [CrossRef] [PubMed]

23. Thakor, N.V.; Pan, K. Ventricular Tachycardia and Fibrillation Detection by a Sequential Hyphesis testing algorithm. *IEEE Trans. Biomed. Eng.* **1990**, *37*, 837–843. [CrossRef] [PubMed]

24. Kuo, S.; Dillman, R. Computer Detection of Ventricular Fibrillation. In Proceedings of the Computers in Cardiology, Los Alamitos, CA, USA, 12–14 September 1978; pp. 347–349.

25. Pan, J.; Tompkins, W. A real-time QRS detection algorithm. *IEEE Trans. Biomed. Eng.* **1985**, *32*, 230–236. [CrossRef] [PubMed]

26. Zhang, X.; Jiang, H.; Zhang, L.; Zhang, C.; Wang, Z.; Chen, X. An Energy-Efficient ASIC for Wireless Body Sensor Networks in Medical Applications. *IEEE Trans. Biomed. Circuits Syst.* **2010**, *4*, 11–18. [CrossRef] [PubMed]

27. Simon, D. Kalman Filtering. *Embed. Syst. Program.* **2001**, *14*, 72–79.

28. Texas Instrument. *ez430-2500 Development Tool—User's Guide, Literature Number: SLAU227E*; Texas Instrument: Dallas, TX, USA, 2009.

29. Thakor, N.V. From Holter Monitors to Automatic Defibrillators: Developments in Ambulatory Arrhythmia Monitoring. *IEEE Trans. Biomed. Eng.* **1984**, *31*, 770–778. [CrossRef] [PubMed]

30. Bradley, A.P. The Use of the Area under the ROC Curve in the Evaluation of Machine Learning Algorithms. *Pattern Recognit.* **1997**, *30*, 1145–1159. [CrossRef]

Article

A Data Compression Hardware Accelerator Enabling Long-Term Biosignal Monitoring Based on Ultra-Low Power IoT Platforms

Christos P. Antonopoulos * and Nikolaos S. Voros

Computer & Informatics Engineering Department, Technological Educational Institute of Western Greece, 30020 Antirio, Greece; voros@teiwest.gr
* Correspondence: cantonopoulos@teiwest.gr; Tel.: +30-693-610-6468

Received: 31 May 2017; Accepted: 27 July 2017; Published: 31 July 2017

Abstract: For highly demanding scenarios such as continuous bio-signal monitoring, transmitting excessive volumes of data wirelessly comprises one of the most critical challenges. This is due to the resource limitations posed by typical hardware and communication technologies. Driven by such shortcomings, this paper aims at addressing the respective deficiencies. The main axes of this work include (a) data compression, and (b) the presentation of a complete, efficient and practical hardware accelerator design able to be integrated in any Internet of Things (IoT) platform for addressing critical challenges of data compression. On one hand, the developed algorithm is presented and evaluated on software, exhibiting significant benefits compared to respective competition. On the other hand, the algorithm is fully implemented on hardware providing a further proof of concept regarding the implementation feasibility with respect to state-of-the art hardware design approaches. Finally, system-level performance benefits, regarding data transmission delay and energy saving, are highlighted, taking into consideration the characteristics of prominent IoT platforms. Concluding, this paper presents a holistic approach based on data compression that is able to drastically enhance an IoT platform's performance and tackle efficiently a notorious challenge of highly demanding IoT applications such as real-time bio-signal monitoring.

Keywords: IoT Wireless Sensor Network platforms; data compression; hardware accelerator; Wireless Sensor Networks; embedded systems; complete solution; experimental evaluation; hardware design; ultra-low power

---

1. Introduction

Internet of Things (IoT) short-range, ultra-low power communication technologies comprise one of the most rapidly evolving research areas attracting significant interest both from academia and the industry [1]. Consequently, respective communication protocols as well as platforms have emerged as prominent infrastructure upon which future Cyber-Physical Systems (CPS) can be based. A major factor for this growth can be attributed to the novel communication paradigm, through which relative communication approaches have been introduced enabling flexible communication, distributed operation and rapid deployment without the need for any pre-existing infrastructure, mobility support, low-power functionality and many more critical features. Additionally, IoT and CPS have significantly benefited from the advancements in hardware and Very Large Scale Integration (VLSI) design leading to very low cost, complexity, size and, most importantly, power consumption embedded systems able to be used as a suitable hardware infrastructure [2–4]. Furthermore, another aspect related to hardware design, concerns the anticipated advantages yielded by the hardware accelerators. The latter, comprise highly specialized hardware components integrated on general-purpose IoT platforms, dedicated to perform highly complex operations in the most efficient way. Such approach offers multifaceted benefits since the design of nodes can exhibit significantly higher degree of dynamicity and modularity, tailored to specific applications. At the same time, demanding operations are performed with maximum efficiency, while the main processing unit of the nodes is not overwhelmed.

A consequence of the aforementioned benefits is that the relative application domain has significantly expanded from simple data processing and transmission scenarios to increasingly demanding ones. Respective examples entail significant processing workload (e.g., medical applications), the support of complex communication standards (e.g., IPv6 routing protocol support enabling true IoT deployment), and the cooperative functionality among nodes in order to achieve a common goal etc. Of course, aspects such as time-constrained communication and power aware operation impose stringent requirements. In order to meet the respective requirements, researchers must surpass the challenge of the extremely limited resource availability encountered in all nowadays Wireless Sensor Networks (WSN) platforms. The respective limitation ranges from low processing power offered by typical low-end micro-controller units, to limited communication bandwidth offered by protocols like IEEE 802.15.4 or Bluetooth, as well as scarce availability in terms of memory (few tens of Kbytes) and energy. Concerning the latter limitation (i.e., energy availability), emphasis must be put on the fact that the wireless radio interface comprises the main power consumption factor, which highlights the respective deficiencies. Therefore, a common critical objective in various pragmatic approaches, aiming towards lifetime extension, is the deactivation of the relative hardware when respective functionality is not required for extended periods of time [5,6].

In this paper, the authors tackle the aforementioned deficiencies stemming from scarce bandwidth and energy availability, which in conjunction with demanding applications comprise a notorious challenge. As it will be analyzed in the following section, real-time analysis of biomedical modalities requires excessive volume of data transmission, forming two challenging side effects. On one hand, it is easily quantifiable that in many cases the bandwidth offered cannot handle the required data rate. On the other hand, even in cases where the channel capacity is adequate for data creation rates, the respective scenarios lead to continuous operation of highly power consuming components, and especially the wireless radio interface. Respective cases contradict typical WSN platform paradigm, which effectively dictates the transition of power-hungry hardware to low-power states for substantial periods of time.

Driven by such conditions, this paper proposes a holistic solution that allows mitigating the respective side effects based on a highly efficient and resource conservative data compression hardware accelerator. Specifically, the critical contribution of this work is the design, development and system level evaluation of the potential benefits of the proposed hardware component taking into consideration realistic, demanding biomedical datasets as well as real performance characteristics of state-of-the-art IoT platform. As it will be presented, the developed compression algorithm and its hardware design is able to compress the targeted modalities on-the-fly. Additionally, it is able to yield a considerable compression rate in all cases, thus enhancing the transmission capabilities of an IoT platform. Furthermore, the proposed hardware accelerator leads to drastic power conservation due to the data volume reduction, assuming that IoT nodes are able to deactivate radio interface when not transmitting. To the best of the authors' knowledge, this is the first attempt to offer a practical, efficient and feasible complete IoT compression component going beyond proposing an isolated software compression approach. On the contrary, the proposed solution has been implemented in hardware and evaluated taking into account realistic wireless transmission delays and power consumption statistics of prominent IoT wireless communication interfaces.

The rest of the paper is structured as follows: In Section 2 critical information is provided highlighting the application scenarios, relative literature and the data characteristics of the specific problem. Section 3 presents the design and evaluation of the proposed compression algorithm yielding the required characteristics so as to optimally address the targeted problem. Section 4 comprises the cornerstone of this work and presents the actual hardware module design, implementation and evaluation. Section 5 offers a critical system level performance evaluation of the proposed hardware accelerator assuming that it has been integrated in prominent real IoT platforms and two different wireless communication interfaces. Finally, Section 6 offers a summarizing discussion on performance evaluation and module's implementation capabilities while Section 7 highlights the main points of this work and offers significant future extensions based on the presented work.

2. Background Information

2.1. Relative WSN Compression Approaches

Based on the previous analysis, an elicitation concerning adequate compression algorithms for epilepsy monitoring and utilization in WSN networks is conducted. The elicitation process followed a multifaceted approach considering (a) the nature of the application scenario the authors focused on, which is electrocardiogram/electroencephalogram (ECG/EEG) physiological signal monitoring, (b) the representation of the digitized data as time-series and (c) the performance aspects taking into consideration the utilization in on-the-fly WSN scenarios. With respect to the first characteristic, the criticality and the accuracy required for the monitoring and study of epileptic people dictates zero tolerance to datum corruption due to the compression process. Consequently, only lossless compression algorithms suitable for wireless sensors have been considered [7,8] as opposed to lossy approaches. Regarding the second parameter, we focused on algorithms well known for their effectiveness on time-series datasets [7–11] taking into consideration that both Electroencephalography (EEG) and Electrocardiography (ECG) result into time-series datasets. Finally, emphasizing on the utilization of such algorithms in the context of WSNs, our elicitation process focuses on low complexity in order to offer viable solutions for typical WSN nodes. In addition, it aims at minimizing the delay overhead and operating in a time constrained manner [10]. Taking into account a wide range of adequate compression approaches, the algorithms Lossless Entropy Compression (LEC) [8] and Adaptive Lossless Entropy Compression (ALEC) [10] were finally selected as the starting point for the hardware implementation. LEC is a low complexity lossless entropy compression algorithm resulting in a very small code footprint that requires very low computational power [12]. Its operation is based on utilizing a very small dictionary and exhibits impressive on-the-fly compression capabilities. Consequently, LEC is quite attractive for WSN deployments. The main steps of the algorithm include:

- Calculation of differential signal.
- Computing of the difference d_i between the binary representations r_i and $r_i - 1$ of the current and previous measurements respectively, encoding is applied upon d_i resulting in the corresponding bit sequence bs_i.
- The sequence bs_i is then concatenated to the bit sequence stream generated so far.

The main processing complexity is attributed to the encoding phase of the compression, which aims at transforming d_i to bs_i bit sequences. During this process, firstly the number n_i of bits needed to encode the value of d_i is computed. Secondly, the first part of bs_i, indicated as s_i, is generated by using the table that contains the dictionary adopted by the entropy compressor. In that respect, in our initial evaluation JPEG algorithm has been adopted because the coefficients used in the JPEG have similar statistical characteristics to the measurements acquired by the sensing unit. In our implementation, the table has been extended so as to cover the necessary 16-bit resolution of the Analog to Digital Convertor (ADC) in the sensors. Thirdly, the second part of bs_i, indicated as a_i is the n_i low-order bits of d_i is calculated [8].

Focusing on the implementation features of LEC, it can be implemented by maintaining in memory only the column s_i of the aforementioned table. Overall, it can be easily extracted that LEC avoids any computationally intensive operation, which is highly appreciated in scarce resource WSN platforms. As a result, it exhibits very low execution delay facilitating real-time operation. However, basing its operation solely on a static lookup table, it does not support dynamic configuration capabilities with respect to the characteristics of a specific signal. The latter, is a significant drawback that negatively affects LEC's compression rate capabilities [12].

ALEC, on the other hand, is based on an adaptive lossless entropy compression approach also requiring low computational power. However, it uses three small dictionaries, the sizes of which are determined by the resolution of the analog-to-digital converter (ADC). Adaptive compression schemes allow the compression to dynamically adjust to the data source. The data sequences to be compressed are partitioned into blocks and for each block the optimal compression scheme is applied. Effectively, the algorithm is similar to LEC algorithm, while the main difference is that ALEC algorithm uses three Huffman coding tables instead of the one table used for the DC coefficients in JPEG algorithm. Specifically, ALEC uses adaptively two Huffman tables and three Huffman tables, respectively. It is noted that compared to LEC, ALEC exhibits an increased number of the lookup tables it employs. As a result, its compression-rate efficiency is also increased. However, each data block is passed through two lookup tables, which eventually will result in an increase of the algorithm's processing delay [12].

2.2. Application Scenarios and Data Characteristics

A critical objective of this effort concerns the performance enhancement offered by an efficient compression module to the state-of-the-art WSN platforms targeting highly demanding applications. Epilepsy monitoring based on IoT platform, represents the main targeted application scenario of this paper. From an engineering point of view, the respective application scenarios are based on acquiring excessive amounts of data (digitized physiological measurements), for extended periods of time, which must be either stored locally or transmitted to an aggregation point. The most demanding case, comprising the main objective of this paper, concerns the real-time monitoring of an epileptic person. This represents a critical requirement since epileptic events can occur unexpectedly and unpredictably, while the triggering event is highly personalized to the specific person. For that reason, periodic-based monitoring leads to myopic and unreliable results and conclusions. However, in both cases highly accurate signal monitoring is also required, which represents also a critical requirement towards effective compression.

WSNs comprise a rapidly evolving research area proposing a new communication paradigm making them appealing for such cases. However, in realistic scenarios the respective platforms suffer from critical resource limitations especially in areas such as energy availability, communication bandwidth and processing power that, at the same time, are interdependent resource consuming factors. More specifically, the operation of state-of-the-art WSN platforms is based on small batteries, typically offering energy capacity from 450 mAh (e.g., Shimmer platform) up to 3300 mAh (e.g., two AA batteries). Consequently, all components must offer low-power characteristics resulting in processing units with low processing capabilities (provided by 16-bit-based Micro-Controller Units) and limited available memory (in the area of 10 Kbyte RAM). Additionally, the data transfer is based on low-power wireless interfaces such as IEEE 802.15.4 and Bluetooth offering bandwidth capabilities at physical layer from 250 Kbps up to a few Mbps.

Taking into consideration the aforementioned application scenarios, the two modalities of paramount importance in epileptic seizure study, resulting into significant amount of accumulated data, are EEG and ECG measurements. Typical acquisition devices produce samples represented as 16-bit numbers. Furthermore, a wired EEG setup is usually comprised of 64 sensors with sampling frequency up to 2.5 kHz, while ECG typically requires 4 sensors with adequate sampling frequency of a few hundreds of Hertz. Given that the main application scenario targeted is the real-time monitoring of epileptic persons, even with some rough calculations, it can be derived that a setup of 64 EEG sensors requires higher wireless bandwidth (not considering packet headers and control data) than the one prominent communication technologies can provide. Specifically, assuming typical wireless technologies such as IEEE 802.15.4 and Bluetooth, such scenarios pose overwhelming burden to WSN platforms typically offering extremely limited resources [9,13–18].

Such workload scenarios demand that power-hungry components (such as the wireless interfaces or/and the processing unit) are continuously operating at their maximum power consumption operational state. Such behavior, however, contradicts to the main objective of typical WSNs of residing such components to low-power states for extended periods of time enabling drastic extension of the network lifetime and thus meet demanding requirements of many hours or days of monitoring.

Therefore, in such cases the reduction of the amount of data (i.e., by compressing them) can yield significant and multifaceted benefits to the system performance. On one hand, if we assume that a specific amount of data is compressed by x%, the respective data volume reduction can be correlated to analogous reduction of radio utilization resulting in significant energy conservation. On the other hand, the x% compression percentage can also be envisioned as analogous reduction of bandwidth requirement and thus effectively increasing the limited wireless channel utilization while reducing the resulting data transfer delay.

3. Compression Algorithm Design

In this section, the proposed, and implemented, lossless entropy compression algorithm is presented, focusing on low computational power. Its main characteristic is the exploitation of the observed data's frequencies in the time series in order to offer an efficient dynamically adaptive Huffman table for the encoding process. The main novelty is based on extending the approaches derived from previously presented LEC and ALEC algorithms (i.e., low complexity, low code footprint, low compression delay). In this way, the proposed algorithm retains the very good processing delay performance characteristic. However, at the same time it significantly enhances achieved compression rate by increasing adaptability to the data characteristics. Consequently, compression rate exhibited by the proposed extension is highly competitive compared to respective solutions published in relative literature, while it also offers optimum compression rate vs compression processing delay trade-off. It is noted that a detailed analysis of the characteristics of this compression approach and evaluation at software level based on Matlab implementations can be found in [12].

3.1. Rationale of the Proposed Algorithm

LEC algorithm compresses data in its entropy encoder with the use of a fixed table leading to a very fast execution performance. This table is an extension of the table used in the JPEG algorithm to reach the size necessary for the resolution of the ADC in use. Additionally, it is based on the fact that the closer the absolute of a value is to zero more frequently it is observed in the differential signal.

However, it has been noticed that this frequency distribution may be valid for a file or a stream of data, however it is not always accurate considering fractions of the file or the data stream. Based on this observation ALEC algorithm uses a small amount of fixed Huffman Codes tables that can be alternatively used to produce smaller code for a packet of data. Furthermore, the specific table is not optimal for the specific data under test at each particular experiment.

Therefore, in the proposed scheme (as depicted in Figure 1) a novel approach is introduced, where the Huffman Codes lookup tables used are continuously adjusted according to the characteristics of the specific data used. Also the degree that the tables are adjusted is also configurable, offering fine tuning capabilities and enhancing the added value of the respective novel approach.

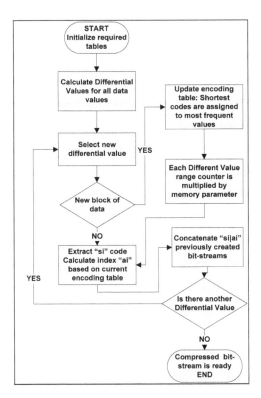

Figure 1. Real-time huffman scheme flow chart.

3.2. Utilization of Data Statistical Knowledge

Usually statistical knowledge is available when time-series data are measured and transmitted through wireless sensors. A method can be used based on earlier observations, but since data are changing over time this knowledge can be of questionable value as far as the compression effectiveness is concerned.

Therefore, in the proposed scheme, the previously observed frequency values are exploited to effectively update Huffman Code tables for the values to follow. Initially, the differential signal is produced, as most likely it has values that can be more effectively compressed. Following the initial phase, the differential signal is separated in fixed size packets. In the first packet, since there is no statistical knowledge of the data, the method is using the table from the LEC algorithm. However, in each following packet the statistical knowledge from previous data is used to create on the fly an adaptive Huffman Code table. The alphabet of numbers is divided into groups, the sizes of which increase exponentially. Each new differential signal value observed, leads to the appropriate group's frequency increment. When the processing of a data packet ends, the frequencies of each group are used to extract the possibilities of a value that belongs to that group. The blocks are sorted in descending order by their possibilities and a binary tree is created.

After the formation of the Huffman Code table for the following packet, the current frequency table is element wise multiplied with a factor varying between 1 and 0. Therefore, as this parameter is approaching 0 the degree by which history (i.e., frequencies observed in previous packets) is taken into account in the next Huffman Code table diminishes. Therefore, if "0" is selected only the frequencies of the cur-rent packet are used; if 1 is selected the frequencies of every previous packet are equally used in the encoding of the next packets [12].

3.3. Performance Analysis of the Proposed Algorithm

In this section, a brief comparative analysis is presented of the proposed compression algorithm (Real Time Huffman) against LEC and ALEC approaches highlighting critical advantages offered. Considering the former comparison, all three algorithms are implemented in Matlab environment and are being evaluated with respect to the compression rate achieved and compression delay assuming execution on the same personal computer. The evaluation is based upon real EEG and ECG signals, extracted either from respective open data bases or from real measurements undertaken in the context of the specific evaluation effort, which comprise the most challenging modalities in epilepsy monitoring. A brief description of the sources of the datasets as well as the execution environment is as follows.

3.3.1. PhysioNet Database

PhysioNet [19] was established in 1999 as the outreach component of the Research Resource for Complex Physiologic Signals cooperative project. From this database signals were extracted and used as evaluation testbeds of the implemented algorithms from the following two specific subcategories:

1. *Apnea-ECG Database*

This database has been assembled for the PhysioNet/Computers in Cardiology Challenge 2000 [20]. From this database the ecgA04apnea and ecgB05apnea used in the evaluation process have been acquired.

2. *CHB-MIT Scalp EEG Database*

This database [21], collected at the Children's Hospital Bos-ton Massachusetts Institute of Technology (MIT), consists of EEG recordings from pediatric subjects with intractable seizures [22].

3.3.2. University of Patras, EEG and ECG Signals

The dataset is provided by the Neurophysiology Unit, Laboratory of Physiology School of Medicine, University of Patras (UoP) [23]. EEG was recorded using 58 EEG Ag-AgCl electrodes according to the extended international 10–20 system.

3.3.3. EkgMove ECG Signals

EkgMove [24] is a psycho-physiological measurement sys-tem for research applications. From the various measurement capabilities of the sensor, the ECG signal of a single subject has been used.

As a result, from the respective evaluation, Figure 2 presents the achieved compression rate of the three different compression algorithms over the dataset of eight different signals is presented. As depicted, the proposed RT-Huffman algorithm manages to offer the highest compression in all cases. Specifically, compared to LEC, RT-Huffman offers an increased compression rate varying from 1.5 up to 4.3% while the same variation with respect to ALEC reaches up to 3.5%.

Table 1 presents the measurements concerning the processing delay of compression algorithms under evaluation. In this case, due to drastic processing delay differences of the different biosignals, in order to extract more objective conclusions Table 1 depicts the absolute delay for the algorithm offering the lowest measurement and the percentage deviation for the other two algorithms.

As extracted from Table 1, LEC yields the lowest processing demands upon the processing unit. However, what is even more important is that in all cases the proposed RT-Huffman proves to be the second less resource demanding solution clearly outperforming ALEC and exhibiting a steady and relative small overhead compared to LEC thus able to meet time constrained performance demands.

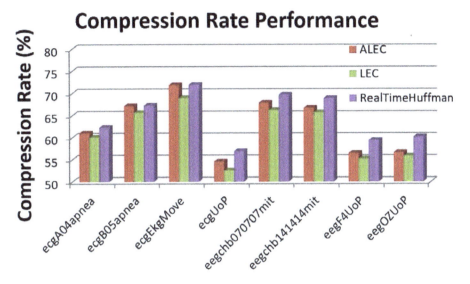

Figure 2. Compression rate performance evaluation.

Table 1. Processing delay performance evaluation.

Data Set	RT-Huffman	LEC (s)	ALEC
ecgA04apnea	+18%	7.5	+30%
ecgB05apnea	+28%	6.7	+35%
ecgEkgMove	+26%	16.8	+35%
ecgUoP	+11%	201	+131%
eegchb070707mit	+14%	18.1	+34%
eegchb141414mit	+23%	17.8	+38%
eegF4UoP	+18%	191.9	+144%
eegOZUoP	+10%	198.17	+124%

4. Hardware Implementation Design

The proposed algorithm was implemented at hardware level and a respective system level performance evaluation was carried out. The compression module processes input samples on the fly with a latency of 4 clock cycles. As depicted in Figure 3, a small 16×16 bit *Input Buffer* stores the incoming 16-bit samples and propagates them to the compression module when instructed by the *Local Controller*. On system power-up the samples are propagated through the differential datapath comprised by a *Subtractor* and an *Absolute Calculation Unit*. The absolute value of all samples is used to update a metric table with statistical information and is also used to produce the compressed output. This is the initialization phase of the system.

When a number of samples equal to the defined block size has been collected, the controller enters in calculations' phase and pauses further sample propagation to the rest of the system. The *Huffman Micro-Processor Unit* calculates and produces the Huffman (S) table, based on the populated metric table, which will be applied on the next block of incoming data. The custom microprocessor functions with encoded operations are designed so as to optimize this phase. The core of the Huffman algorithm is implemented by performing parallel memory accesses on the 33×9 bits *Tableful Parallel Memory* and by un-rolling all nested while and for loops to serial operations. A 512×36 *Instruction Memory* drives the micro-processor to execute all Real Time Huffman algorithm calculations which leads to a substantial processing latency. The worst-case processing latency due to the iterative nature of the algorithm is calculated to 4175 clock cycles per block of 500 samples. Once the Huffman (S) table has been calculated, the controller resumes propagation of samples through the differential data-path and S is applied to the next block of incoming samples in order to produce the compressed output. When a number of samples equal to the defined block size has been collected, the calculation phase is activated again and so on.

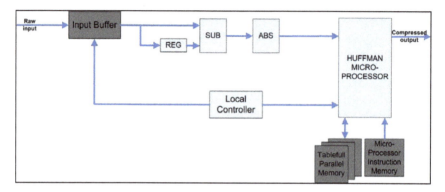

Figure 3. Compression module block diagram.

In Figure 4, the block data flow with respect to compression of the input data block and the calculations of the S table is depicted. It is noted (indicated with the immediate receipt of the data of the 4th data block following the 3rd data block) that if the intermediate interval between the last datum of one block and the first datum of the subsequent is smaller than the time required to perform the S table calculations, the Ready For Data (RFD) signal will go low; then excessive data can be stored in an input buffer until the S table calculations are concluded, and then the compression process resumes.

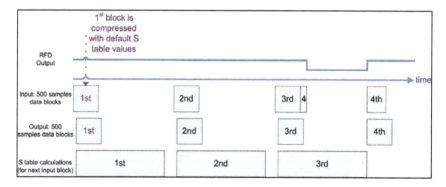

Figure 4. Compression module block data flow.

4.1. I/Os Timing Protocols

As seen in Figure 5, the mechanism has a 4 CPU cycle delay between the time an uncompressed datum is processed and the time the corresponding compressed equivalent is ready at the output. Therefore, and based on the lookup table at that point, datum $0 \times \text{FFD8}$ corresponds to $0 \times 39\text{A}$ requiring 10 bits according to the lookup table, datum 0×000 corresponds to $0 \times 3\text{A}5$ also corresponding to 10 bits and so on.

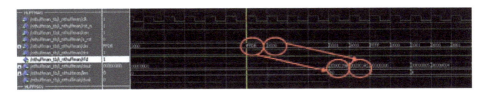

Figure 5. Compression module Input/Output block timing. Start of block.

The same timing process is followed until the end of the block as indicated in Figure 6, where the cycled datum is the 500th (i.e., last) datum of the block corresponding to the corresponding cycled datum of data out pin.

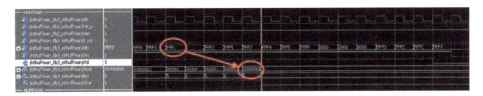

Figure 6. Compression module Input/Output block timing. End of block.

4.2. Implementation Results

In Table 2, the processing performance characteristics are presented regarding the processing delay of one block of data and the corresponding throughput considering a relative low frequency clock, adequate for embedded systems.

Table 2. Compression module processing delay and throughput rate (Assuming × MHz operating frequency).

Processing Delay (In Clock Cycles)	Throughput Rate (Worst Case)
4175 (Worst Case)	106.8 Mbps (62.5 MHz Clk)

Calculation of the throughput rate:
A data block of 500, 16 bit samples will be processed at worst case every:
4 cc (data-path latency) + 500 cc (input samples) + 4175 c (processing delay) = 4679 cc.

$$\text{Throughput rate (bps)} = (16 \times 500 \text{ bits})/(4679 \times \text{Tclk}), \tag{1}$$

For 62.5 MHz CLK, we get worst case processing latency estimation at 106.8 Mbps, which is equal to 13.375 "500 × 16 bits samples" data-blocks per second. Based on such throughput capacity the following comparison with software-based respective implementation can be made.

Considering 10-minute EEG signals acquired from the MIT data base [22] the corresponding file is 307.200 bytes and the required compression delay exhibited by the software-based algorithm was measured to be approximately 20 s considering an i7 dual core pc. Based on the previous calculation, the respective delay of the hardware accelerator is anticipated to be around 20 ms, thus offering a drastic decrease regarding delay as well as resource consumption. Furthermore, such evaluation indicates that the presented hardware accelerator comprises a highly efficient solution when considering multiple channels acquired concurrently or/and low frequency processing units being available by the sensors.

Additionally, the implementation presented is resource conservative as indicated in Table 3, depicting the degree of FPGA resources required by the specific implementation. Such measurements effectively indicate that more than one hardware accelerator could coexist in the same FPGA board offering increased flexibility.

Table 3. Compression module resources (Spartan3E xc3s100 – 4ft256 Xilinx FPGA Technology).

Configurable Logic Block (CLB) Slices	Flip Flops (FFs)	Block RAMs	I/Os
1676/7680 – 21%	1148/15,360 – 7%	7/12 – 29%	60/173 – 34%

4.3. Power Dissipation

For 1 Kbps requirement, 1 kHz clock will be sufficient. However due to XPower Analyzer limitations [23] we can get a power estimate for frequencies as low as 1 MHz (for worst case 100% toggling rates). Specifically, the total power usage of an FPGA device can be broken down to total static power and total dynamic power. Static power is associated with DC current while dynamic power is associated with AC current. Static power is power consumed while there is no circuit activity. Dynamic power is power consumed while there is inputs/circuit activity (Table 4).

Table 4. Compression module power dissipation (Spartan3E Xilinx FPGA Technology).

Clock Frequency (MHz)	Total Power (mW)	Dynamic Power (mW)	Static Power (mW)
1	114.27	15.53	98.74
62.5	198.52	98.45	100.07

5. System-Level Performance Evaluation

The main objective of this section is to analyze both qualitatively and quantitatively the benefits that the integration of the proposed compression module can offer to a real WSN platform, by taking into account both the performance and the energy conservation.

The degree by which the bandwidth required to transmit a specific amount of data is reduced, due to utilization of compression, depends solely on the algorithm and not on the integration efficiency. However, the system-level transmission delay and the energy consumption metrics depend heavily on the hardware design of the compression modules, as well as on the integration of the component to a specific WSN platform. To evaluate both compression delay and compression power consumption, we assume a signal sample of X Bytes that can be compressed by Y% and adopt the following rationale.

As far as the transmission delay is concerned, the two equations that enable us to calculate the required time interval to send the respective information with and without the compression module are the following ones.

Without the compression module integrated:

$$Tx_{no\_compress} = Tx_{XBytes}, \tag{2}$$

where Tx_{XBytes}: Represents the time interval required to successfully transmit X Bytes and it is measured using a specific platform and specific WSN communication technology.

With the compression module integrated:

$$Tx_{compress} = Tc_{XBytes} + Tx_{x \times y\% Bytes} + Tdc_{XBytes}, \tag{3}$$

where Tc_{XBytes} is the time interval required to compress X Bytes that can be accurately calculated for a specific processing clock frequency of the implementation presented in Section 4. $Tx_{x \times y\% Bytes}$ is the time required to successfully transmit the compressed amount of data which can be accurately measured using a specific platform and a specific WSN communication technology. Tdc_{XBytes} is the time interval required to decompress the compressed data back to X Bytes which, without loss of accuracy, can be considered equal to Tc_{XBytes} since the compression and decompression algorithms are symmetrical in the proposed solution.

Moving on to energy consumption the respective equations used are as follows.

Energy consumption without integrated compression module:

$$Ex_{no\_compress} = P_{tx\_radio} \times Tx_{XBytes} = V_{tx\_radio} \times I_{tx\_radio} \times Tx_{XBytes}, \tag{4}$$

where $Ex_{no\_compress}$ is the energy consumed for sending successfully X Bytes of data. P_{tx_radio} is the transmission power consumption of a specific radio transceiver. V_{tx_radio} is the voltage supply required of specific radio transceiver (provided by respective datasheets). I_{tx_radio} is the current draw during transmission of a specific radio transceiver (provided by respective datasheets). Tx_{XBytes} is the time required to successfully transmit X bytes; it is measured using a specific platform and a specific communication technology.

Energy consumption with integrated compression module:

$$Ex_{compress} = Ec_{XBytes} + E_{tx_x \times y\%Bytes} = (P_{compr} \times Tc_{XBytes}) + (P_{tx_{radio}} \times Tx_{X \times y\%Bytes}) =$$
$$(P_{compr} \times Tc_{XBytes}) + (V_{tx\_radio} \times I_{tx\_radio} \times Tx_{X \times y\%Bytes}), \tag{5}$$

where $Ex_{compress}$ is the energy required to compress X Bytes and successfully transmit the compressed data. Ec_{XBytes} is the energy required for compressing X Bytes. $E_{tx_x \times y\%Bytes}$ is the energy required to successfully transmit the compressed data. P_{compr} is the power consumption (static + dynamic) of the compression module, when active. Tc_{XBytes} is time interval required to compress X Bytes, which can be accurately calculated for a specific processing clock frequency. V_{tx_radio} is the voltage supply required for a specific radio transceiver (provided by respective datasheets). I_{tx_radio} is the current draw during the transmission of specific radio transceiver (provided by respective datasheets). $Tx_{X \times y\%Bytes}$ is the time required to successfully transmit $X \times Y\%$ Bytes; it is measured using a specific platform and a specific WSN communication technology.

It is noted that energy consumption is not considered for the decompression phase, under the assumption that the receiver (i.e., the home gateway) is always plugged in the main power supply. Furthermore, two assumptions are made so as to extract the results presented in following section. First, it is assumed that when the compression module is not used it is shut down and does not consume energy; second, it is assumed that when the radio does not transmit it is tuned off and thus there is no energy consumption.

5.1. Performance Evaluation Setup

In order for the analysis approach presented in the previous section to lead to realistic and useful results, it is important to present the main hardware components (including the different configuration options considered) utilized as well as the actual data, comprised by samples taken from real EEG/ECG signals, which are two of the most demanding modalities in epilepsy monitoring (which represents our main application scenario).

The most important component influencing our evaluation is the implemented compression hardware module. As analyzed in Section 4, the hardware module throughput (thus the delay required to compress a specific amount of data) depends on the clock frequency based on Equation (1).

The power consumption depends also on the frequency clock. Therefore, we considered four different clock frequencies; then, we measured the power consumption and based on Equation (5) the following table has been extracted. Table 5 presents the two main performance metrics of the compression hardware implementation proposed.

Table 5. Power consumption and throughput performance of compression hardware component.

Clock Frequency (MHz)	Total Power (mW)	Dynamic Power (mW)	Static Power (mW)	Throughput (Mbps)
1	114.27	15.53	98.74	17.09
20	142.38	43.20	99.18	34.19
40	170.07	70.45	99.62	68.39
62.5	198.52	98.45	100.07	106.86

Table 5 depicts also a tradeoff that, as will be proven, comprises a critical factor with respect to the overall performance evaluation. As depicted, as the clock frequency increases, both power consumption and throughput also increase. Consequently, when compressing a specific amount of data increasing the clock frequency implies that, on one hand, the component needs more energy while active but, on the other hand, it performs its operation much faster, and consequently the time period required to be active is drastically reduced. A secondary observation concerns the fact that the main factor of power consumption increase is the dynamic power and not the static power (its influence is rather negligible).

5.1.1. Short-Range Wireless Communication Radio Characteristics

The second component influencing the system-level performance is the specific off-the-shelf radio transceivers used to transfer data. In the context of this evaluation, we considered the Shimmer2R platform [25] comprising a prominent commercial solution offering highly efficient Bluetooth Class 2.1 as well as IEEE 802.15.4 interfaces.

Bluetooth Interface Characteristics:
The specific platform uses the RN-42 Bluetooth radio chip exhibiting the following voltage and current demands [26].
Voltage supply = 3.3 V
Typical Current draw during transmission = 45 mA
Additionally, the respective node is programmed using TinyOS operating system so additional delay could be added as expected to be the case in any off-the-shelf WSN sensor.

IEEE 802.15.4 Interface Characteristics:
The specific platform uses the CC2420 IEEE 802.15.4 radio chip exhibiting the following voltage/current demands [14].
Voltage supply = 3.3 V
Typical Current draw during transmission (max) = 17.4 mA

Additionally, the respective node is programmed using TinyOS operating system so additional delay could be added as expected to be the case in any off-the-shelf WSN sensor.

5.1.2. Dataset Compression Characteristics

The actual signals comprising the evaluation dataset comprise also a significant parameter influencing the performance of hardware accelerator. It is assumed that all datasets are parts of signals of equal time duration and specifically of 10min interval. Specifically, the signals considered are derived from the extended dataset presented in Section 3.3 and the characteristics relevant to the specific evaluation are as follows:

1. ECG_UoP

An ECG signal with sampling rate 2.5 KHz, 16 bits per sample and 10 min duration leads to an amount of data equal to 3.000.000 Bytes; the implemented compression algorithm is able to compress them to 1.292.952 Bytes, thus yielding 56% compression rate. Shimmer Platform using BT technology is able to transfer compressed and uncompressed data in 60 and 25 s respectively. Shimmer Platform using IEEE 802.15.4 technology is able to transfer compressed and uncompressed data in 488 and 211 s respectively considering a configuration of 108 byte per packet and a packet transmitter every 2 ms.

2. ECG_EkgMove

An ECG signal acquired using the EkgMove sensor with sampling rate of 256 Hz, 16 bits per sample and 10 min duration leads to an amount of data equal to 307.200 Bytes; the implemented compression algorithm is able to compress to 86.536 Bytes yielding 72% compression rate. Shimmer Platform using BT technology is able to transfer compressed and uncompressed data in 6 and 2 s respectively. Shimmer Platform using IEEE 802.15.4 technology is able to transfer compressed and uncompressed data in 50 and 15 s respectively.

5.2. Evaluation Results

In this section, we present the most interesting quantitative results of our evaluation. In order to extract the respective observations more easily, Figure 7 presents the results from Tables 6 and 7 in comparison to the power demands of the compression module.

Table 6. Energy and delay reduction for ECG_UoP using shimmer bluetooth nodes.

Clock	Transmission Delay (s)			Energy Consumption (mJ)		
MHz	Without Compression	With Compression	% Reduction	Without Compression	With Compression	% Reduction
1	60	53.074	11.5	8910	5316.5	40.3
20	60	26.4	55.99	8910	3812.4	57.2
40	60	25.7	57.1	8910	3772.18	57.8
62.5	60	25.44	57.5	8910	3757.1	57.6

Table 7. Energy and delay reduction for ECG_UoP using shimmer 802.15.4 nodes.

Clock	Transmission Delay (s)			Energy Consumption (mJ)		
MHz	Without Compression	With Compression	% Reduction	Without Compression	With Compression	% Reduction
1	488	239	51	28,020.96	13,719.63	51
20	488	212	56	28,020.96	12,215.55	56
40	488	211	57	28,020.96	12,175.3	57
62.5	488	211	57	28,020.96	12,160.21	57

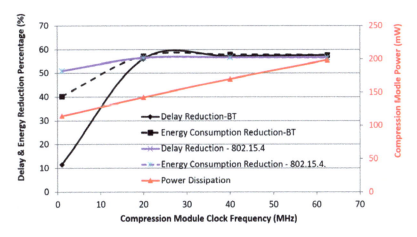

Figure 7. Compression module performance effect wrt ECG_UoP/shimmer setup.

5.2.1. ECG_UoP

Based on Equations (2)–(5) and the individual components' characteristics as presented in previous section, the analytical results are presented in Tables 6 and 7.

As it can be easily extracted, significant benefits can be offered by integrating the proposed compression module into a WSN node. What is interesting to note is that when increasing the frequency clock, although it leads to increased power for the compression module, it also leads to performance enhancements in both metrics considered i.e., end-to-end delay and power consumption. This is attributed to the considerably faster operation of the module, which decreases drastically the time period required by the module to be active. However, this performance enhancement also seems to have an upper limit, since increasing the frequency beyond 20 MHz it leads to negligible performance enhancement. Thus, the respective increase to power requirements beyond 20 MHz clock frequency is effectively wasted since it does not lead to further performance enhancement.

Finally, it is important to emphasize the fact that a reduction of almost 60% can be calculated both for the delay and the energy dissipation using the proposed hardware compression module.

Focusing on IEEE 802.15.4-based measurements, what is especially important in this case is that a drastic enhancement is apparent even with the lowest clock frequency considered (i.e., only 1 MHz). This is attributed to the much longer delay required to transmit data compared to the BT case. Consequently, in this case, it proves significantly beneficial to compress data before wireless transmission. This is indicated both delay and energy wise, since the compression phase, due to its hardware implementation, leads to drastically low delay operation, which in turn minimizes any end-to-end delay and energy overhead. Furthermore, the same pattern is identified in the graphs indicating 20 MHz as the optimum frequency selection, yielding considerable performance enhancement, and limited power consumption overhead. Overall, this maximum performance enhancement is a bit lower compared to BT case but still reaching up to 57%.

5.2.2. ECG_EkgMove

Based on Equations (2)–(5) the analytical results are as follows described in Tables 8 and 9. In order to extract the respective observations more easily, in Figure 8 the results from Tables 8 and 9 are presented in comparison to the power demands of the compression module.

Table 8. Energy and delay reduction for ECG_EkgMove using shimmer BT nodes.

Clock	Transmission Delay (s)			Energy Consumption (mJ)		
MHz	Without Compression	With Compression	% Reduction	Without Compression	With Compression	% Reduction
1	6	4.87	19	891	461.25	48
20	6	2.14	64	891	307.23	65
40	6	2.07	66	891	303.11	66
62.5	6	2.04	65	891	301.56	66

Table 9. Energy and delay reduction for ECG_EkgMove using shimmer 802.15.4 nodes.

Clock	Transmission Delay (s)			Energy Consumption (mJ)		
MHz	Without Compression	With Compression	% Reduction	Without Compression	With Compression	% Reduction
1	50	17.9	64	2871	1025.55	64
20	50	15.1	70	2871	871.53	70
40	50	15.1	70	2871	867.41	70
62.5	50	15.1	70	2871	865.87	70

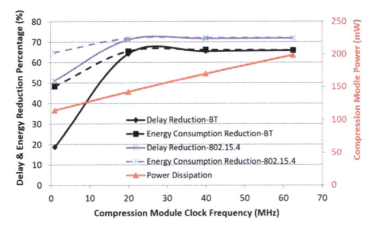

Figure 8. Compression module performance effect wrt ECG_EkgMove/shimmer BT setup.

Figure 8 reveals behavioral patterns similar to the ones in previous cases, strengthening the indications regarding the performance enhancement of the compression module. Focusing on BT-based graph, an interesting point is the fact that in comparison to the respective measurements considering ECG UoP signal, in this case a steadily higher performance enhancement (~8%) can be extracted. This enhancement increase is attributed to the lower sampling frequency of the EkgMove sensor leading to fewer samples for same time interval, thus leading significantly faster compression execution from the proposed compression module. This is actually the part that effectively saves network resources either in the delay or in energy consumption domain. Also, in this case the type of signal has a critical role; this is due to the fact that the more compression prone the signal is the higher the performance enhancement will be, which in this case is close to 70% both for delay and energy reduction.

Focusing on 802.15.4-based graphs, once again the exhibited delay and energy consumption decrease are both quite emphatic. Both reach up to 70%, which can be achieved even with a moderate FPGA clock frequency of 20 MHz (62.5 MHz being the highest supported by the FPGA development board considered for the implementation).

Another interesting observation concerns the fact that even considering the significantly lower power consumption of IEEE 802.15.4 transceivers (compared to Bluetooth counterpart), the integration of a compression accelerator module would still lead to a significant performance boost and consequent enhance the efficiency of any decision support system.

6. Discussion

This paper explores the possibility to exploit data compression, in order to effectively tackle the wireless data transmission requirements of Epilepsy monitoring by typical WSN platforms. Adequate and objective study of Epilepsy requires continuously acquisition of respective modalities at high rates, thus creating excessive volumes of data. Following a multifaceted research and development approach, in this paper, firstly the problem is addressed algorithmically by the comparison of software-based compression algorithms targeting at optimum trade-off between compression rate and compression delay. In that respect, as analytically presented in [12] the proposed algorithm highly competitive performance. Secondly, going a step further towards practical solutions, the proposed algorithm is implemented at hardware level in the context of a typical FPGA platform. The implemented hardware module is presented in detail in this paper, while performance wise it reduced the compression delay by a factor of 1/1000. At the same time, dynamic power consumption ranges between 15 and 98 mW representing a viable solution for contemporary WSN platforms. The latter is further emphasized in the third part of this work, where the anticipated effect of integrating such a component with real prominent WSN platforms is evaluated.

In this aspect two well-known wireless interfaces are considered, characterized by difference features and capabilities. Furthermore, the effect of running the proposed module at different clock frequencies is also taken into consideration. In all cases the benefits of using the proposed solution are quiet apparent. Specifically, considering Bluetooth wireless transmission, end-to-end delay reduction ranges between 11% and 65%, while energy is respective reduction ranges between 40% and 66%. Performance enhancement is even higher when considering IEEE 802.15.4 wireless communication since both transmission delay and energy consumption are reduced by from 5% to 70%.

Without a doubt, such performance and resource conservation enhancements offer convincing arguments that hardware-based data compression comprises a viable solution towards network/node lifetime increase and wireless communication bandwidth optimum management.

7. Conclusions

In order for IoT platforms to comprise a reliable and efficient technological infrastructure for nowadays as well as future CPS applications, novel hardware, real-time approaches are required in order to enhance the platforms' capabilities while minimizing resource wastage. This paper addresses a notorious challenge of wireless interfaces' excessive utilization considering highly demanding application scenarios such as bio-signal monitoring. To tackle respective shortcomings of nowadays platforms, the two main pillars of this paper are (a) the design and implementation of an efficient, real-time compression algorithm, and (b) the design and implementation of a highly efficient and resource conservative hardware accelerator for the proposed algorithm. The idea behind this approach is to offer on-the-fly hardware compression to any IoT platform, thus overloading the main processing unit, which is usually of limited processing capabilities. Consequently, this paper offers a holistic solution to the problem of wirelessly transmitting excessive volumes of data, tackling all aspects from the design and software algorithmic performance, to the design of the actual hardware accelerator and finally the overall system performance. With respect to the latter, taking into consideration the power consumption of a real prominent IoT platform and respective delay transmission capabilities, the proposed hardware compression accelerator can yield an approximate 70% system wide delay and energy consumption reduction. Furthermore, the proposed module comprises a highly feasible and practical solution since it captures just 21% of the Spartan3 FPGA's configurable logic block slices and the maximum performance gains are achieved at just 1/3 of the maximum frequency clock tested. Such characteristics advocate the integration of the proposed module to nowadays WSN platforms.

Acknowledgments: This study is part of the collaborative project armor which is partially funded by the European commission under the seventh framework programme (FP7/2007–2013) with grant agreement: 287720 and through the horizon 2020 programme (H2020/PHC-19-2014) under the research and innovation action "radio-robots in assisted living environments: unobtrusive, efficient, reliable and modular solutions for independent ageing".

Author Contributions: Christos P. Antonopoulos conceived and designed the experiments concerning the proposed compression algorithm. Nikolaos S. Voros conceived and implemented the hardware implementation of the compression algorithm's module. Christos P. Antonopoulos conceived, designed and carried out the integrated experiments, Both Christos P. Antonopoulos and Nikolaos S. Voros contributed on data analysis. Both Christos P. Antonopoulos and Nikolaos S. Voros wrote the paper.

Conflicts of Interest: The authors declare no conflict of interest.

References

1. Perera, C.; Liu, C.H.; Jayawardena, S. The emerging internet of things marketplace from an industrial perspective: A survey. *IEEE Trans. Emerg. Top. Comput.* **2015**, *3*, 585–598. [CrossRef]
2. Zivkovic, M.; Popovic, R.; Tanaskovic, T. The Survey of Commercially Available Sensor Platforms from Energy Efficiency Aspect. In Proceedings of the 20th Telecommunications Forum (TELFOR), Belgrade, Serbia, 20–22 November 2012; pp. 1528–1531.
3. Xu, T.; Wendt, J.B.; Potkonjak, M. Security of IoT Systems: Design Challenges and Opportunities. In Proceedings of the 2014 IEEE/ACM International Conference on Computer-Aided Design, San Jose, CA, USA, 3–6 November 2014; IEEE Press: Piscataway, NJ, USA, 2014. [CrossRef]

4. Tonneau, A.S.; Mitton, N.; Vandaele, J. A Survey on (Mobile) Wireless Sensor Network Experimentation Testbeds. In Proceedings of the 2014 IEEE International Conference on Distributed Computing in Sensor Systems, Marina Del Rey, CA, USA, 26–28 May 2014; pp. 263–268. [CrossRef]

5. Cotuk, H.; Tavli, B.; Bicakci, K.; Akgun, M.B. The Impact of Bandwidth Constraints on the Energy Consumption of Wireless Sensor Networks. In Proceedings of the 2014 IEEE Wireless Communications and Networking Conference, Istanbul, Turkey, 6–9 April 2014; pp. 2787–2792. [CrossRef]

6. Abdel-Aal, M.O.; Shaaban, A.A.; Ramadan, R.A.; Abdel-Meguid, M.Z. Energy Saving and Reliable Data Reduction Techniques for Single and Multi-Modal WSNs. In Proceedings of the 2012 International Conference on Engineering and Technology (ICET), Cairo, Egypt, 10–11 October 2012; pp. 1–8. [CrossRef]

7. Salomon, D. *Data Compression: The Complete Reference*, 4th ed.; Springer Science and Business Media: Berlin, Germany, 2004.

8. Marcelloni, F.; Vecchio, M. An efficient lossless compression algorithm for tiny nodes of monitoring wireless sensor networks. *Comput. J.* **2009**, *52*, 969–987. [CrossRef]

9. Low, K.S.; Win, W.N.N.; Er, M.J. Wireless Sensor Networks for Industrial Environments. In Proceedings of the International Conference on Computational Intelligence for Modelling, Control and Automation and International Conference on Intelligent Agents, Web Technologies and Internet Commerce, Vienna, Austria, 28–30 November 2005; pp. 271–276.

10. Kolo, J.G.; Shanmugam, S.A.; Lim, D.W.G.; Ang, L.M.; Seng, K. P. An adaptive lossless data compression scheme for wireless sensor networks. *J. Sens.* **2012**. [CrossRef]

11. Takezawa, T.; Asakura, K.; Watanabe, T. Lossless compression of time-series data based on increasing average of neighboring signals. *Electron. Commun. Jpn.* **2010**, *93*, 47–56. [CrossRef]

12. Antonopoulos, C.P.; Voros, N.S. Resource efficient data compression algorithms for demanding, WSN based biomedical applications. *Elsevier J. Biomed. Inform.* **2016**, *59*, 1–14. [CrossRef] [PubMed]

13. Antonopoulos, C.; Prayati, A.; Stoyanova, T.; Koulamas, C.; Papadopoulos, G. Experimental Evaluation of a WSN Platform Power Consumption. In Proceedings of the 2009 IEEE International Symposium on Parallel & Distributed Processing, Rome, Italy, 23–29 May 2009; pp. 1–8.

14. Polastre, J.; Szewczyk, R.; Sharp, C.; Culler, D. The Mote Revolution: Low Power Wireless Sensor Network Devices. In Proceedings of the Hot Chips, Stanford, CA, USA, 22–24 August 2004.

15. Ultra Low Power IEEE 802.15.4 Compliant Wireless Sensor Module. Available online: http://www.eecs.harvard.edu/~konrad/References/TinyOSDocs/telos-reva-datasheet-r.pdf (accessed on 30 July 2017).

16. Jurcik, P.; Koubaa, A.; Severino, R.; Alves, M.; Tovar, E. Dimensioning and worst-case analysis of cluster-tree sensor networks. *ACM Trans. Sens. Netw. (TOSN)* **2010**, *7*, 14. [CrossRef]

17. Dietrich, I.; Dressler, F. On the lifetime of wireless sensor networks. *ACM Trans. Sens. Netw. (TOSN)* **2009**, *5*, 5. [CrossRef]

18. Mottola, L.; Picco, G.P.; Ceriotti, M.; Guna, S.; Murphy, A.L. Not all wireless sensor networks are created equal: A comparative study on tunnels. *ACM Trans. Sens. Netw. (TOSN)* **2010**, *7*, 15. [CrossRef]

19. Goldberger, A.L.; Amaral, L.A.N.; Glass, L.; Hausdorff, J.M.; Ivanov, P.C.; Mark, R.G.; Mietus, J.E.; Moody, G.B.; Peng, C.K.; Stanley, H.E. PhysioBank, PhysioToolkit, and PhysioNet: Components of a New Research Resource for Complex Physiologic Signals. *Circulation* **2000**, *101*, e215–e220. [CrossRef] [PubMed]

20. Penzel, T.; Moody, G.B.; Mark, R.G.; Goldberger, A.L.; Peter, J.H. The Apnea-ECG Database. In Proceedings of the Computers in Cardiology 2000, Cambridge, MA, USA, 24–27 September 2000; pp. 255–258.

21. Shoeb, A.H. Application of Machine Learning to Epileptic Seizure Onset Detection and Treatment. Ph.D. Thesis, Massachusetts Institute of Technology, Cambridge, MA, USA, 10 September 2009.

22. PhysioNet. Available online: http://www.physionet.org/ (accessed on 30 July 2017).

23. Xpower Analyzer. Available online: http://www.xilinx.com/support/documentation/sw_manuals/xilinx11/xpa_c_overview.htm (accessed on 30 July 2017).

24. Movisens GmbH. Available online: https://www.movisens.com/en/products/ (accessed on 30 July 2017).

25. Shimmer WSN Platform. Available online: http://www.shimmer-research.com/ (accessed on 30 July 2017).

26. Roving Networks Bluetooth™ Product User Manual. Available online: http://www.mct.net/download/rn/rn-41_42.pdf (accessed on 30 July 2017).

Article

Pipelined Architecture of Multi-Band Spectral Subtraction Algorithm for Speech Enhancement

Mohammed Bahoura

Department of Engineering, Université du Québec à Rimouski, 300, allée des Ursulines, Rimouski, QC G5L 3A1, Canada; Mohammed_Bahoura@uqar.ca; Tel.: +1-418-723-1986

Received: 29 August 2017; Accepted: 27 September 2017; Published: 29 September 2017

Abstract: In this paper, a new pipelined architecture of the multi-band spectral subtraction algorithm has been proposed for real-time speech enhancement. The proposed hardware has been implemented on field programmable gate array (FPGA) device using Xilinx system generator (XSG), high-level programming tool, and Nexys-4 development board. The multi-band algorithm has been developed to reduce the additive colored noise that does not uniformly affect the entire frequency band of useful signal. All the algorithm steps have been successfully implemented on hardware. Pipelining has been employed on this hardware architecture to increase the data throughput. Speech enhancement performances obtained by the hardware architecture are compared to those obtained by MATLAB simulation using simulated and actual noises. The resource utilization, the maximum operating frequency, and power consumption are reported for a low-cost Artix-7 FPGA device.

Keywords: FPGA; hardware/software co-simulation; pipelining; speech enhancement; multi-band spectral subtraction; signal-to-noise ratio

---

1. Introduction

The enhancement of speech corrupted by background noise represents a great challenge for real-word speech processing systems, such as speech recognition, speaker identification, voice coders, hand-free systems, and hearing aids. The main purpose of speech enhancement is to improve the perceptual quality and intelligibility of speech by using various noise reduction algorithms.

Spectral subtraction method is a popular single-channel noise reduction algorithm that has been initially proposed for speech enhancement [1]. This basic spectral subtraction method can substantially reduce the noise level, but it is accompanied by an annoying noise in the enhanced speech signal, named musical noise. A generalized version of this method has been proposed to reduce the residual musical noise by an over-subtraction of the noise power [2]. The multi-band algorithm has been developed to reduce additive colored noise that does not uniformly affect the entire frequency band of useful speech signal [3]. Improved version based on multi-band Bark scale frequency spacing has been also proposed to reduce the colored noise [4]. An adaptive noise estimate for each band has been proposed [5]. Furthermore, the spectral subtraction approach has also been applied to other kinds of sounds such as underwater acoustic sounds [6], machine monitoring [7,8], hearing aid [9], pulmonary sounds [10–12], etc.

In real-world applications, such as hands-free communication kits, cellular phones and hearing aid devices, these speech enhancement techniques need to be executed in real-time. Hardware implementation of this kind of algorithms is a difficult task that consists in finding a balance between complexity, efficiency and throughput of these algorithms. Architectures based on the spectral subtraction approach have been implemented on Field Programmable Gate Array (FPGA) devices [13–19]. However, these architectures perform a uniform spectral subtraction over the entire frequency band and, therefore, they do not efficiently suppress colored noise.

In this paper, a new pipelined architecture of multi-band spectral subtraction method has been proposed for real-time speech enhancement. The proposed architecture has been implemented on FPGA using the Xilinx System Generator (Xilinx Inc, San Diego, CA, USA) programming tool and the Nexys-4 (Digilent Inc, Pullman, WA, USA) development board build around an Artix-7 XC7A100T FPGA chip (Xilinx Inc, San Diego, CA, USA). Mathematical equations describing this speech enhancement algorithm (Fourier transform, signal power spectrum, noise power estimate, multi-band separation, signal-to-noise ratio, over-subtraction factor, spectral subtraction, multi-band merging, inverse Fourier transform, etc.) have been efficiently modeled using the XSG blockset. High-speed performance was obtained by inserting and redistributing the pipelining delays.

The rest of the paper is organized as following: Section 2 presents the theory details of the spectral subtraction theory for speech enhancement. Section 3 presents the XSG-based hardware system and discusses the details of different subsystems. Speech enhancement performances are presented in Section 4. Finally, conclusion and perspective are provided in Section 5.

2. Spectral Subtraction Methods

In the additive noise model, it is assumed that the discrete-time noisy signal $y[n]$ is composed of the clean signal $s[n]$ and the uncorrelated additive noise $d[n]$.

$$y[n] = s[n] + d[n] \qquad (1)$$

where n is the discrete-time index.

Since the Fourier transform is linear, this relation is also additive in the frequency domain

$$Y[k] = \mathcal{F}\{y[n]\} = S[k] + D[k] \qquad (2)$$

where $Y[k]$, $S[k]$ and $D[k]$ are the discrete Fourier transform (DFT) of $y[n]$, $s[n]$ and $d[n]$, respectively, and k is the discrete-frequency index. In practice, the spectral subtraction algorithm operates on a short-time signal, by dividing the noisy speech $y[n]$ into frames of size N. Then, the discrete Fourier transform is applied on each frame. The frequency resolution depends both on the sampling frequency f_S and the length of the frame in samples N. The discrete frequencies are defined by $f_k = kf_S/N$ for $0 \leq k < N/2$, and by $f_k = (k - N)f_S/N$ for $N/2 \leq k < N$. As a complex function, the Fourier transform of the corrupted signal can be represented by its rectangular form $Y[k] = Y_r[k] + jY_i[k]$, where $Y_r[k]$ and $Y_i[k]$ are the real and imaginary part of $Y[k]$, respectively. It can also be represented by its polar form $Y[k] = |Y[k]|e^{j\varphi_y[k]}$, where $|Y[k]|$ and $\varphi_y[k]$ are the magnitude and the phase of $Y[k]$, respectively.

2.1. Basic Spectral Substraction

In the spectral subtraction method, the spectrum of the enhanced speech is obtained by subtracting an estimate of noise spectrum from the noisy signal spectrum [1]. To avoid negative magnitude spectrum, a simple half-wave rectifier has been first employed.

$$|\widehat{S}[k]| = \max \left\{ |Y[k]| - |\widehat{D}[k]|, \ 0 \right\} \qquad (3)$$

where the noise spectrum, $\widehat{D}[k]$, is estimated during the non-speech segments.

To reconstruct the enhanced signal, its phase $\varphi_{\hat{s}}[k]$ is approximated by the phase $\varphi_y[k]$ of the noisy signal. This is based on the fact that in human perception the short time spectral magnitude is more important than the phase [4,20]. Thus, the discrete Fourier transform (DFT) of the enhanced speech is estimated as

$$\widehat{S}[k] = |\widehat{S}[k]| \ e^{j\varphi_y[k]} \qquad (4)$$

Finally, the enhanced speech $\hat{s}[n]$ is obtained by inverse discrete Fourier transform (IDFT).

$$\hat{s}[n] = \mathcal{F}^{-1}\{\hat{S}[k]\} \tag{5}$$

However, this basic method suffers from a perceptually annoying residual noise named musical noise.

2.2. Generalized Spectral Subtraction

A generalized form of the spectral subtraction method has been suggested [2] to minimize the residual musical noise. It consists of over-subtracting an estimate of the noise power spectrum and ensuring that the resulting spectrum does not fall below a predefined minimum level (spectral floor).

$$|\hat{S}[k]|^{\gamma} = \max\left\{|Y[k]|^{\gamma} - \alpha|\hat{D}[k]|^{\gamma}, \ \beta|\hat{D}[k]|^{\gamma}\right\} \tag{6}$$

where $\alpha \geq 1$ is the over-subtraction multiplication factor and $0 < \beta \ll 1$ is the spectral flooring parameter [2]. γ is the exponent determining the transition sharpness, where $\gamma = 1$ corresponds to the magnitude spectral subtraction [1] and $\gamma = 2$ to the power spectral subtraction [2].

To minimize the speech distortion produced by large values of α, it has been proposed to let α vary from frame to frame within speech signal [2].

$$\alpha = \begin{cases} 4.75 & \text{SNR} \leq -5; \\ 4 - \frac{3}{20}\text{SNR} & -5 \leq \text{SNR} \leq 20; \\ 1 & \text{SNR} \geq 20. \end{cases} \tag{7}$$

where SNR is the segmental signal-to-noise ratio estimated in the frame and defined by:

$$\text{SNR} = 10\log_{10}\left(\frac{\sum\limits_{k=0}^{N-1}|Y[k]|^2}{\sum\limits_{k=0}^{N-1}|\hat{D}[k]|^2}\right) \tag{8}$$

As for the basic method, the discrete Fourier transform of the enhanced signal $\hat{S}[k]$ is calculated from the estimated magnitude $|\hat{S}[k]|$ of the enhanced signal and the phase $\varphi_y[k]$ of the corrupted input signal using (4). The enhanced signal $\hat{s}[n]$ is reconstructed by inverse discrete Fourier transform (5).

2.3. Multi-Band Spectral Subtraction

The multi-band spectral subtraction method has been developed to reduce additive colored noise that does not uniformly affect the entire frequency band of the speech signal [3]. In this method, both noisy speech and estimated noise spectra are divided into M non-overlapping frequency bands. Then, the generalized spectral subtraction is applied independently in each band [3]. The power spectrum estimate of the enhanced speech in the ith frequency band is obtained as:

$$|\hat{S}_i[k]|^2 = \max\{|Y_i[k]|^2 - \alpha_i\delta_i|\hat{D}_i[k]|^2, \ \beta|\hat{D}_i[k]|^2\} \qquad b_i \leq k \leq e_i \tag{9}$$

where b_i and e_i are the beginning and the ending frequency bins of the ith frequency band ($1 \leq i \leq M$), α_i is the over-subtraction factor of the ith frequency band, and δ_i is the tweaking factor of the ith frequency band [3]. The over-subtraction factor α_i is related to the segmental SNR_i of the ith frequency band by:

$$\alpha_i = \begin{cases} 4.75 & \text{SNR}_i \leq -5; \\ 4 - \frac{3}{20}\text{SNR}_i & -5 \leq \text{SNR}_i \leq 20; \\ 1 & \text{SNR}_i \geq 20. \end{cases} \tag{10}$$

The segmental SNR_i of the ith frequency band is defined by:

$$\text{SNR}_i = 10\log_{10}\left(\frac{\sum\limits_{k=b_i}^{e_i}|Y_i[k]|^2}{\sum\limits_{k=b_i}^{e_i}|\hat{D}_i[k]|^2}\right) \tag{11}$$

where b_i and e_i are the beginning and the ending frequency bins of the ith frequency band. It can also be expressed using the natural logarithmic function:

$$\text{SNR}_i = \frac{10}{\ln(10)}\left(\ln\left(\sum_{k=b_i}^{e_i}|Y_i[k]|^2\right) - \ln\left(\sum_{k=b_i}^{e_i}|\hat{D}_i[k]|^2\right)\right) \tag{12}$$

The tweaking factor δ_i in (9) can be used to have an additional degree of noise removing control in each frequency band. The values of δ_i are experimentally defined and set to [3]:

$$\delta_i = \begin{cases} 1 & f_i \leq 1 \text{ kHz}; \\ 2.5 & 1 \text{ kHz} < f_i \leq 2 \text{ kHz}; \\ 1.5 & 2 \text{ kHz} < f_i \leq 4 \text{ kHz}. \end{cases} \tag{13}$$

where f_i denotes frequency in the ith band.

Figure 1 shows the block diagram of the speech enhancement system based on the multi-band spectral subtraction approach. Input noisy speech, $y[n]$, is segmented into consecutive frames of N samples before applying discrete Fourier transform (DFT). The magnitude, $|Y[k]|$, and phase, $\varphi_y[k]$, of the Fourier transformed signal are calculated. Then, spectrum of the noisy speech, $|Y[k]|^2$, is calculated for the current frame, while the spectrum of the noise, $|\hat{D}[k]|^2$, is estimated during non-speech segments. Both spectra are separated into ($M = 4$) frequency bands of 1 kHz in width each, for a sampling frequency $f_s = 8$ kHz. The segmental SNR_i and the over-subtraction factor α_i are calculated for each frequency band to allow independent spectral subtraction. The 4 separate spectra of the enhanced speech, $|\hat{S}_i[k]|^2$, are then merged and square root calculated to obtain $|\hat{S}[k]|$. Finally, the enhanced signal $\hat{s}[n]$ is reconstructed by using inverse discrete Fourier transform (IDFT).

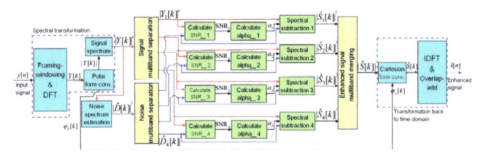

Figure 1. Block diagram of the multi-band spectral subtraction for speech enhancement. DFT: discrete Fourier transform; IDFT: inverse discrete Fourier transform.

3. FPGA Implementation

The proposed architecture has been implemented on a low-cost Artix-7 FPGA chip using a high-level programming tool (Xilinx System Generator), in MATLAB/SIMULINK (The Mathworks Inc., Natick, MA, USA) environment, and Nexys-4 development board. The top-level Simulink diagram of this architecture is presented in Figure 2, which principally corresponds to the block diagram presented in Figure 1. The proposed architecture uses some subsystems (blocks) developed in the past to implement, on an FPGA chip, the basic spectral subtraction method [15]. However, only subsystems related to the multi-band approach will be described in details in the following subsections:

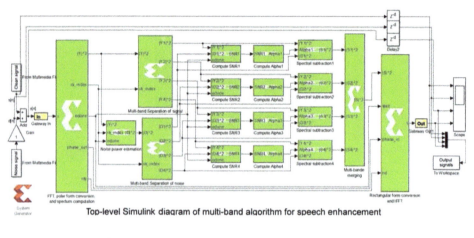

Top-level Simulink diagram of multi-band algorithm for speech enhancement

Signal-to-noise (SNR) calculation

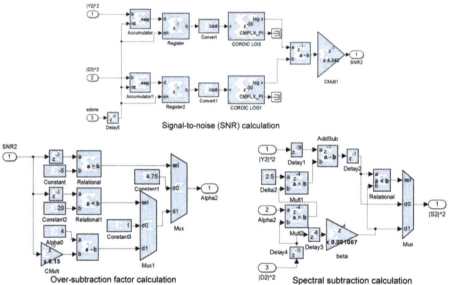

Over-subtraction factor calculation

Spectral subtraction calculation

Figure 2. The proposed pipelined architecture of multi-band spectral subtraction using Xilinx system generator (XSG) blockset. The top-level Simulink diagram is shown on the top followed by details of the main calculation subsystems (signal-to-noise ratio, over-subtraction factor, and spectral subtraction). Details about the implementation of noise spectrum estimate can be found in [15].

3.1. Spectral Transformation and Noise Spectrum Estimation

This step corresponds to the four left blue blocks in Figure 1 and the two left XSG subsystems in Figure 2. As described in [15], the spectral analysis is performed by a Xilinx FFT (Fast Fourier Transform) block that provides the real, $Y_r[k]$, and imaginary, $Y_i[k]$, parts of the transformed signal $Y[k]$. It also provides the frequency index k output and an output *done* indicating that the computation of the Fourier transform of the current frame is complete and ready to output. Then, the Xilinx CORDIC (COordinate Rotation DIgital Computer) block is used to convert the transformed signal to its polar form, i.e., magnitude ($|Y[k]|$) and phase ($\varphi_y[k]$). A simple multiplier is used to calculate the power spectrum, $|Y[k]|^2$, of the noisy signal.

On the other hand, the power spectrum, $|\widehat{D}[k]|^2$, of the additive noise is estimated using its average value calculated during the first five frames. A RAM (Random Access Memory)-based accumulator is used to estimate the noise power. More details on this subsystem can be found in [15].

3.2. Multi-Band Separation of Signal and Noise

This step corresponds to the two left yellow blocks in Figure 1 and their associated two subsystems in Figure 2. The hardware implementation of this subsystem is done using four register (one per frequency band) having as input the signal power spectrum, $|Y[k]|^2$, and driven by the frequency index signal, k. If k belongs to the ith frequency band, then the ith register is enabled by k, i.e., $|Y_i[k]|^2=|Y[k]|^2$. Otherwise, it is reset, i.e., $|Y_i[k]|^2 = 0$. For the ith band, the frequency index k is delimited by b_i and e_i, as in (11) and (12). The same subsystem is used to separate the noise power spectrum into linearly separated multi-band.

3.3. Signal-To-Noise (SNR) Estimator

This step corresponds to the four left green blocks in Figure 1 and their associated four subsystems in Figure 2. Considering the fact that Xilinx CORDIC block can calculate only the natural logarithm (ln) and not the decimal logarithm (log) function, the SNR subsystem has been implemented using (12) instead of (11). In addition, this approach permits avoiding the use of divider.

As shown in Figure 2, this subsystem uses accumulators and registers to compute the sums for both signal and noise, followed by the CORDIC blocks to calculate their respective ln functions. After the subtractor block, the resulting value is multiplied by constant $4.3429 = 10/\ln(10)$.

3.4. Over-Subtraction Factor Calculation

This step corresponds to the four middle green blocks in Figure 1 and their associated four subsystems in Figure 2. Based on (10), this subsystem is implemented using two multiplexers, two comparators, one subtractor, and constant blocks. The subsystem used in the 2nd frequency band is shown in Figure 2.

3.5. Spectral Subtraction

This step corresponds to the four right green blocks in Figure 1 and their associated four subsystems in Figure 2. This subsystem is implemented using one comparator, one multiplexer, and three multipliers, according to (9). Figure 2 shows the subsystem used in the 2nd frequency band.

3.6. Multi-Band Merging

This step corresponds to the right yellow blocks in Figure 1 and its associated subsystem in Figure 2. The subsystem is implemented using tree Xilinx adder blocks to merge the spectra of the four sub-bands.

3.7. Transformation Back to Time-Domain

This step corresponds to the two right blue blocks in Figure 1 and the right XSG subsystem in Figure 2. The Fourier transform of the enhanced signal $S[k]$ is first converted to the rectangular form (to real, $S_r[k]$, and imaginary, $S_i[k]$, parts) using a Xilinx CORDIC block, then transformed back to time domain ($s[k]$) using XSG IFFT (Inverse Fast Fourier Transform) block. More details on this subsystem can be found in [15].

3.8. Pipelining

The pipelining consists in reducing the critical path delay by inserting delays into the computational elements (multipliers and adders) to increase the operating frequency (Figure 2). The critical path corresponds to the longest computation time among all paths that contain zero delays [21]. About 410 delays have been inserted in different paths and balanced to ensure synchronization. The proposed pipelining increased substantiality the operating frequency, but at the cost of the output latency of 36 samples. More details on how delays are inserted and redistributed can be found in our previous works [21,22].

3.9. Implementation Characteristics

Table 1 shows the hardware resources utilization, the maximum operating frequency, and the power consumption for the Artix-7 XC7A100T FPGA chip, as reported by Xilinx ISE 14.7 tool (Xilinx Inc, San Diego, CA, USA). The proposed pipelined architecture consumes 4955 logic slices from 15,850 available on this chip (31.2%). Also, it consumes 59 DSP48E1s from 240 available (24.6%). It occupies a small part of this low-cost FPGA. Therefore, the used resources consume about 107 mW. It can be noted that the pipelining of the implemented architecture increased the operating frequency from 24 MHz to 125 MHz, at the cost of an output latency of 36 delays. Therefore, the pipelined architecture requires more flip-flops (20,020 instead of 16,287) because of the inserted delays. Both default and pipelined architectures use the same number DSP48E1s and RAMB18E1s.

Table 1. Resource utilization, maximum operating frequency, and total power consumption obtained for the Artix-7 XC7A100T chip. LUT: Look-Up Table; IOB: Input/Output Block.

Architecture	Default	Pipelined
Resource utilization		
Slices (15,850)	4,617 (29.1%)	4,955 (31.2%)
Flip Flops (126,800)	16,287 (12.8%)	20,020 (15.8%)
LUTs (63,400)	15,067 (23.7%)	16,541 (26.1%)
Bonded IOBs (210)	42 (20.0%)	42 (20.0%)
RAMB18E1s (270)	6 (2.2%)	6 (2.2%)
DSP48E1s (240)	59 (24.6%)	59 (24.6%)
Maximum Operating Frequency	24.166 MHz	125.094 MHz
Total power consumption	107 mW	107 mW

It can be noted that a 32-bit fixed-point format with 28 fractional bits has been globally used to quantify data. However, a 24-bit fixed-point format with 18 fractional bits has been sufficient to compute the over-subtraction factor α_i and the segmental SNR_i.

3.10. Hardware/Software Co-Simulation

Software simulation of the designed XSG-based model provides a faithful behavior to that performed on hardware (bit and cycle accurate modeling). This allows us to take full advantage of the simulation environment of SIMULINK to visualize intermediate signals and facilitate the tuning of the XSG block parameters in order to reach the desired performance. It also optimizes resources by choosing the number of bits needed to quantify data in different paths that ensures the needed performances. However, the designed XSG-based architecture can be executed on actual FPGA chip using the hardware-in-the-loop co-simulation from MATLAB/SIMULINK environment [23]. A number of development boards are pre-configured on the XSG tool, but the Nexys-4 board is not included and must be configured manually. This compilation mode generates a bitstream file and its associate gray SIMULINK block (Figure 3).

During the hardware/software co-simulation, the compiled model (bitstream file) is uploaded and executed on the FPGA chip from SIMULINK environment. XSG tool takes data from the input wav files in SIMULINK environment and transmits them to the design on the FPGA board using the JTAG (Joint Test Action Group) connection. It reads the enhanced signal (output) back from JTAG and sends it to SIMULINK for storage or display.

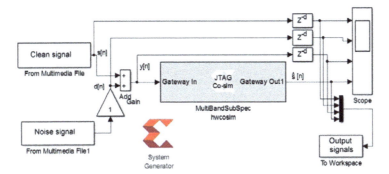

Figure 3. Hardware co-simulation corresponding to the diagram of Figure 2, where $s[n]$ is clean speech signal, $d[n]$ is the noise, $y[n]$ is the corrupted speech signal, and $\hat{s}[n]$ is the enhanced speech. JTAG: Joint Test Action Group protocol.

4. Results and Discussion

The proposed architecture has been tested on hardware using natural speech corrupted by artificial and actual additional noises, with sampling frequency of 8 kHz. The main objective of our experimental test is to validate the implementation process by ensuring that the hardware architecture (XSG) gives the same enhancement performances than the software simulation (MATLAB). Comparison of the multi-band spectral subtraction algorithm to other speech enhancement methods has been evaluated in the literature.

Figures 4 and 5 show the enhancement performances for speech corrupted by artificial blue and pink noises, respectively. However, Figures 6 and 7 present the enhancement performances for speech corrupted by actual car and jet noises, respectively. The fixed-point XSG implementation of the multi-band spectral subtraction technique performs as well as the floating-point MATLAB simulation. Waveforms and spectrograms of the speech signals enhanced by hardware and software are similar. The experimental tests prove the accuracy of the FPGA-based implementation. It can be noted that noise was estimated during the first five frames.

On the other hand, the spectrograms of Figures 4–7 show that the additive noises are removed despite the fact that they do not uniformly affect the frequency band of the speech. For each figure, the difference (error) between the signal enhanced by XSG-based architecture and the signal enhanced by MATLAB simulation has been represented with the same scale in the time-domain. Its time-frequency characteristics seem be close to those of a white noise (quantization noise).

The enhancement performances of the proposed XSG-based architecture are also compared to those obtained by MATLAB simulation using two objective tests: the overall signal-to-noise ratio (oveSNR) and the segmental signal-to-noise ratio (segSNR).

The overall signal-to-noise ratio (oveSNR) of the enhanced speech signal is defined by:

$$\text{oveSNR}_{\text{dB}} = 10 \log_{10} \left(\frac{\sum_{n=0}^{L-1} s^2[n]}{\sum_{n=0}^{L-1} (\hat{s}[n] - s[n])^2} \right) \quad (14)$$

where $s[n]$ and $\hat{s}[n]$ are the original and enhanced speech signals, respectively, and L is the length of the entire signal in samples.

The segmental signal-to-noise ratio (segSNR) is calculated by averaging the frame-based SNRs over the signal.

$$\text{segSNR}_{\text{dB}} = \frac{1}{M} \sum_{m=0}^{M-1} 10 \log_{10} \left(\frac{\sum_{n=0}^{L_s-1} s^2[m,n]}{\sum_{n=0}^{L_s-1} (\hat{s}[m,n] - s[m,n])^2} \right) \quad (15)$$

where M is the number of frames, L_s is the frame size, and $s[m,n]$ and $\hat{s}[m,n]$ are the m-th frame of original and enhanced speech signals, respectively.

Table 2 presents the objective tests for noisy and enhanced speech signals. The results obtained by the proposed XSG-based architecture are approximately similar to those obtained by MATLAB simulation. The minor differences between oveSNR and segSNR can be explained by the quantification errors.

Table 2. Objective tests obtained for noisy and enhanced speech signals for various additive noises. XSG: Xilinx system generator; SNR: segmental signal-to-noise ratio

Objective Test	Blue Noise			Pink Noise			Volvo Car Noise			F-16 Jet Noise		
	Noisy	MATLAB	XSG	Noisy	MATLAB	XSG	Noisy	MATLAB	XSG	Noisy	MATLAB	XSG
oveSNR (dB)	11.92	16.34	16.24	8.57	11.96	11.95	5.04	12.83	13.84	6.18	9.94	9.79
SegSNR (dB)	1.77	8.66	8.00	−2.10	3.95	3.74	−5.27	7.32	7.73	−4.72	2.39	2.08

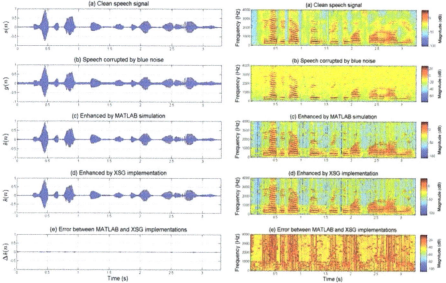

Figure 4. Time waveforms (**left**) and spectrograms (**right**) of clean speech signal (**a**), speech signal corrupted with artificial blue noise (**b**), enhanced speech with floating-point MATLAB simulation (**c**), enhanced with fixed-point XSG implementation (**d**), and error (difference) enhancement between MATLAB and XSG implementations (**e**).

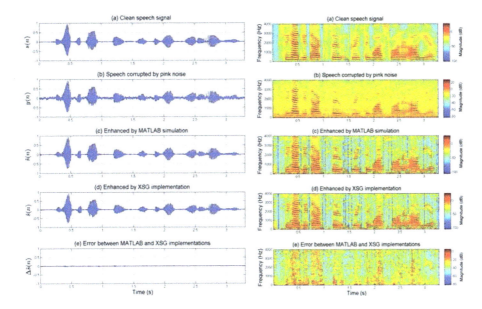

Figure 5. As in Figure 4 but for pink noise.

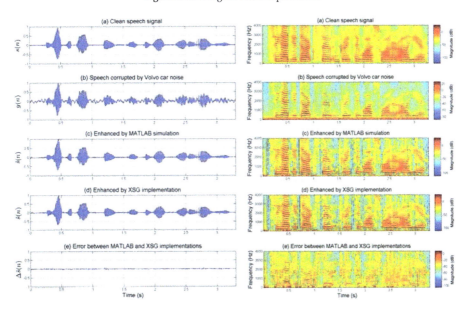

Figure 6. As in Figure 4 but for Volvo car noise.

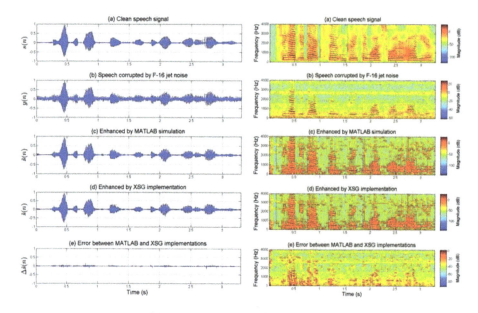

Figure 7. As in Figure 4 but for F-16 jet noise.

5. Conclusions

A pipelined architecture of multi-band spectral subtraction for speech enhancement has been implemented on a low-cost FPGA chip. It occupies a small part of the hardware resources (about 30% of the logic slices) and consumes a low power of 107 mW. The proposed pipelined architecture is five times faster the the default implementation (operating frequency of 125 MHz rather than 24 MHz). Performances obtained by the proposed architecture on speech corrupted by artificial and actual noise are similar to those obtained by MATLAB simulation. For each frequency band, the over-subtraction parameter is adjusted by the signal-to-noise ratio (SNR) to preserve the low-power speech components. However, in the future, a voice activity detector can be added to this architecture in order to update continuously the noise spectrum estimate.

Acknowledgments: This research is financially supported by the Natural Sciences and Engineering Research Council (NSERC) of Canada.

Conflicts of Interest: The author declare no conflict of interest.

References

1. Boll, S.F. Suppression of Acoustic Noise in Speech Using Spectral Subtraction. *IEEE Trans. Acoust. Speech Signal Process.* **1979**, *27*, 113–120.
2. Berouti, M.; Schwartz, R.; Makhoul, J. Enhancement of speech corrupted by acoustic noise. In Proceedings of the IEEE International Conference on ICASSP 1979 Acoustics, Speech, and Signal Processing, Washington, DC, USA, 2–4 April 1979; Volume 4, pp. 208–211.
3. Kamath, S.; Loizou, P. A multi-band spectral subtraction method for enhancing speech corrupted by colored noise. In Proceedings of the IEEE International Conference on Acoustics, Speech, and Signal Processing, Orlando, FL, USA, 13–17 May 2002; Volume 4, pp. 4160–4164.
4. Udrea, R.M.; Vizireanu, N.; Ciochina, S.; Halunga, S. Nonlinear spectral subtraction method for colored noise reduction using multi-band Bark scale. *Signal Process.* **2008**, *88*, 1299–1303.
5. Upadhyay, N.; Karmakar, A. An Improved Multi-Band Spectral Subtraction Algorithm for Enhancing Speech in Various Noise Environments. *Procedia Eng.* **2013**, *64*, 312–321.

6. Simard, Y.; Bahoura, M.; Roy, N. Acoustic Detection and Localization of whales in Bay of Fundy and St. Lawrence Estuary Critical Habitats. *Can. Acoust.* **2004**, *32*, 107–116.

7. Dron, J.; Bolaers, F.; Rasolofondraibe, I. Improvement of the sensitivity of the scalar indicators (crest factor, kurtosis) using a de-noising method by spectral subtraction: Application to the detection of defects in ball bearings. *J. Sound Vib.* **2004**, *270*, 61–73.

8. Bouchikhi, E.H.E.; Choqueuse, V.; Benbouzid, M.E.H. Current Frequency Spectral Subtraction and Its Contribution to Induction Machines' Bearings Condition Monitoring. *IEEE Trans. Energy Convers.* **2013**, *28*, 135–144.

9. Yang, L.P.; Fu, Q.J. Spectral subtraction-based speech enhancement for cochlear implant patients in background noise. *Acoust. Soc. Am. J.* **2005**, *117*, 1001–1004.

10. Karunajeewa, A.; Abeyratne, U.; Hukins, C. Silence-breathing-snore classification from snore-related sounds. *Physiol. Meas.* **2008**, *29*, 227–243.

11. Chang, G.C.; Lai, Y.F. Performance evaluation and enhancement of lung sound recognition system in two real noisy environments. *Comput. Methods Programs Biomed.* **2010**, *97*, 141–150.

12. Emmanouilidou, D.; McCollum, E.D.; Park, D.E.; Elhilali, M. Adaptive Noise Suppression of Pediatric Lung Auscultations With Real Applications to Noisy Clinical Settings in Developing Countries. *IEEE Trans. Biomed. Eng.* **2015**, *62*, 2279–2288.

13. Whittington, J.; Deo, K.; Kleinschmidt, T.; Mason, M. FPGA implementation of spectral subtraction for in-car speech enhancement and recognition. In Proceedings of the 2nd International Conference on Signal Processing and Communication Systems, ICSPCS 2008, Gold Coast, Australia, 15–17 December 2008; pp. 1–8.

14. Mahbub, U.; Rahman, T.; Rashid, A.B.M.H. FPGA implementation of real time acoustic noise suppression by spectral subtraction using dynamic moving average method. In Proceedings of the IEEE Symposium on Industrial Electronics and Applications, ISIEA 2009, Kuala Lumpur, Malaysia, 4–6 October 2009; Volume 1, pp. 365–370.

15. Bahoura, M.; Ezzaidi, H. Implementation of spectral subtraction method on FPGA using high-level programming tool. In Proceedings of the 24th International Conference on Microelectronics (ICM), Algiers, Algeria, 16–20 December 2012; pp. 1–4.

16. Adiono, T.; Purwita, A.; Haryadi, R.; Mareta, R.; Priandana, E. A hardware-software co-design for a real-time spectral subtraction based noise cancellation system. In Proceedings of the 2013 International Symposium on Intelligent Signal Processing and Communications Systems (ISPACS), Naha, Japan, 12–15 November 2013; pp. 5–10.

17. Kasim, M.; Adiono, T.; Fahreza, M.; Zakiy, M. Real-time Architecture and FPGA Implementation of Adaptive General Spectral Substraction Method. *Procedia Technol.* **2013**, *11*, 191–198.

18. Oukherfellah, M.; Bahoura, M. FPGA implementation of voice activity detector for efficient speech enhancement. In Proceedings of the IEEE 12th International New Circuits and Systems Conference, Trois-Rivieres, QC, Canada, 22–25 June 2014; pp. 301–304.

19. Amornwongpeeti, S.; Ono, N.; Ekpanyapong, M. Design of FPGA-based rapid prototype spectral subtraction for hands-free speech applications. In Proceedings of the 2014 Asia-Pacific Signal and Information Processing Association Annual Summit and Conference (APSIPA), Siem Reap, Cambodia, 19–22 December 2014; pp. 1–6.

20. Wang, D.; Lim, J. The unimportance of phase in speech enhancement. *IEEE Trans. Acoust. Speech Signal Process.* **1982**, *30*, 679–681.

21. Bahoura, M.; Ezzaidi, H. FPGA-Implementation of Parallel and Sequential Architectures for Adaptive Noise Cancelation. *Circ. Syst. Signal Process.* **2011**, *30*, 1521–1548.

22. Bahoura, M. FPGA implementation of high-speed neural network for power amplifier behavioral modeling. *Analog Integr. Circ. Signal Process.* **2014**, *79*, 507–527.

23. Bahoura, M. FPGA Implementation of Blue Whale Calls Classifier Using High-Level Programming Tool. *Electronics* **2016**, *5*, 8, doi:10.3390/electronics5010008.

 *electronics*

Article

Real-Time and High-Accuracy Arctangent Computation Using CORDIC and Fast Magnitude Estimation

Luca Pilato, Luca Fanucci and Sergio Saponara *

Dipartimento Ingegneria della Informazione, Università di Pisa, via G. Caruso 16, 56122 Pisa, Italy; luca.pilato@for.unipi.it (L.P.); luca.fanucci@unipi.it (L.F.)
* Correspondence: sergio.saponara@unipi.it; Tel.: +39-050-221-7511; Fax: +39-050-221-7522

Academic Editors: Christos Koulamas and Mihai Lazarescu
Received: 11 February 2017; Accepted: 13 March 2017; Published: 16 March 2017

Abstract: This paper presents an improved VLSI (Very Large Scale of Integration) architecture for real-time and high-accuracy computation of trigonometric functions with fixed-point arithmetic, particularly arctangent using CORDIC (Coordinate Rotation Digital Computer) and fast magnitude estimation. The standard CORDIC implementation suffers of a loss of accuracy when the magnitude of the input vector becomes small. Using a fast magnitude estimator before running the standard algorithm, a pre-processing magnification is implemented, shifting the input coordinates by a proper factor. The entire architecture does not use a multiplier, it uses only shift and add primitives as the original CORDIC, and it does not change the data path precision of the CORDIC core. A bit-true case study is presented showing a reduction of the maximum phase error from 414 LSB (angle error of 0.6355 rad) to 4 LSB (angle error of 0.0061 rad), with small overheads of complexity and speed. Implementation of the new architecture in 0.18 μm CMOS technology allows for real-time and low-power processing of CORDIC and arctangent, which are key functions in many embedded DSP systems. The proposed macrocell has been verified by integration in a system-on-chip, called SENSASIP (Sensor Application Specific Instruction-set Processor), for position sensor signal processing in automotive measurement applications.

Keywords: real-time; Digital Signal Processing (DSP); Embedded Systems; CORDIC (Coordinate Rotation Digital Computer); ASIC (Application Specific Integrated Circuit); FPGA (Field Programmable Gate Array); IP (Intellectual Property); automotive sensors

1. Introduction

The CORDIC (Coordinate Rotation Digital Computer) algorithm [1] is used in many embedded signal-processing applications, where data has to be processed with a trigonometric computation. Rotary encoders, positioning systems, kinematics computation, phase trackers, coordinate transformation and rotation are only a few examples [2–4]. One of those is the use of CORDIC as an arctangent computation block, from here referenced as atan-CORDIC. For example, it is at the core of the phase calculation of complex-envelope signals from the in-phase and quadrature components in communication systems, or for angular position measurement in automotive applications, or for phase computation in power systems control and in AC circuit analysis [5–9]. Although the standard CORDIC algorithm is well known in literature, its efficient implementation in terms of computation accuracy and with low overheads in terms of circuit complexity and power consumption is still an open issue [6–16]. To this aim, both full-custom (ASIC in [6]) and semi-custom (IP macrocells synthesizable on FPGA or standard-cell technologies in [8,9,16]) approaches have been followed in the literature. Due to the need for flexibility in modern embedded systems, the design of a synthesizable IP is proposed in this work.

The iterative algorithm of atan-CORDIC, even if implemented in an IPA (Infinite Precision Arithmetic), suffers from approximation errors due to the n finite iteration. Using the ArcTangent Radix (ATR) angles set, and using vectors compliant with the CORDIC convergence theorem [10], the phase error E_θ is bounded by the expression in Equation (1), which depends on the number of iterations n. This means that ideally we can reduce the angle computation error using more iterations of the CORDIC algorithm.

$$E_\theta \leq \tan^{-1}(2^{-n+1}) \tag{1}$$

The most accurate way to approximate IPA is by relying on floating-point computation, e.g., using the IEEE 754 double-precision floating-point format. However, when targeting real-time and low-complexity/low-power implementations, the design of fixed-point macrocells is more suited. Moreover, many applications using CORDIC, such as position sensor conditioning on-board vehicles, typically do not adopt floating-point units. For example, in automotive or industrial applications, at maximum 32-bit fixed-point processors are used [5,17].

The implementation of the fixed-point algorithm in a Finite Precision Arithmetic (FPA) implies an accuracy degradation of the phase calculation. The contribution to the errors in FPA is the quantization of the ATR angles set and the quantization on the x, y data path, where rounding or truncation action is taken after the shift operation in each step. Using a similar approach to [11,12], we can estimate a coarse upper-bound of the phase error in FPA, as shown in Equation (2).

$$E_\theta \leq \sin^{-1}(2^{-n+1} + \frac{\varepsilon}{r}\sqrt{2}(n-1)) + n\delta \tag{2}$$

In Equation (2), n is the number of iterations, ε is the fixed-point precision of the coordinates' data path where the quantization is taken at each loop, r is the magnitude of the starting point and δ is the angle data path precision. If the initial vector has a magnitude r great enough to neglect the second argument of the $\sin^{-1}$ function in Equation (2), then the error bound depends mainly on the number of iterations. Due to the quantization in the data paths of the fixed-point implementation, every loop of the algorithm has a truncation error that accumulates itself. This is the meaning of the terms $(n-1)$ and $n\delta$ in Equation (2). Particularly, the latter can lead to an increase of the error upper bound in each iteration. Instead, the 2^{-n+1} term indicates the improvement of the error due to more iterations. An example of the dependence of the error upper bound on the number n, once the parameters ε, r and δ are fixed, will be shown in Section 3. It must be noted that Equation (2) expresses an upper bound to the error, and not the real error (see the difference between the red lines and blue lines in Figures 6–9 in Section 3).

For vectors with a magnitude near zero, the error bound and the estimated error increase ideally up to $\pi/2$ due the $\sin^{-1}$ term in Equation (2). The idea proposed in this work is to have a fast estimation of the magnitude of the initial vector and to design a circuit able to magnify the coordinates to let the standard atan-CORDIC work in the area where the phase error is small.

Hereafter, Section 2 presents the fast magnitude estimator. Section 3 discusses the achieved performance in terms of the computation accuracy of the new algorithm. Section 4 discusses the performance of the IP macrocell implementation in CMOS (Complementary Metal Oxide Semiconductor) technology, and its integration in an automotive system-on-chip for sensor signal processing, called SENSASIP (Sensor Application Specific Instruction-set Processor). A comparison of the achieved results to the state-of-the-art in terms of processing accuracy, circuit complexity, power consumption and processing speed is also shown. Conclusions are drawn in Section 5.

2. Fast Magnitude Estimator and Improved CORDIC Architecture

The circuit for a fast magnitude estimation is based on the alphamax-betamin technique [13]. The magnitude estimation is a linear combination of the max and min absolute values of the input coordinates, as shown in Equation (3). The choice of *a* and *b* values depends on the metric that the approximation in Equation (3) wants to minimize. If the max absolute error of the entire input space is the selected metric, the optimum numerical values are $a \approx 0.9551$ and $b \approx 0.4142$. A coarse approximation for a simplified and fast implementation can be $a = 1$ and $b = 0.5$. This means that only an abs block, a comparison block and a virtual shift plus addition have to be implemented in hardware, while avoiding the use of multipliers. Using the architecture in Figure 1, which implements the calculation in Equation (3), the estimated magnitude m^* is always higher than the real one. For example, if the coordinates are equal, i.e., $|x_0| = |y_0|$ the result of Equation (3) is $m* = |x_0| \cdot 3/2$ while the real magnitude is $m_{real} = |x_0| \cdot \sqrt{2} = m * \cdot 2\sqrt{2}/3$. In the case of $|x_0| = 0\ then\ m_{real} = m* = |y_0|$, while if $|y_0| = 0\ then\ m_{real} = m* = |x_0|$. These considerations are shown in Figure 2, where the light grey circumference arc includes the possible positions of the points with the estimated magnitude m^*. Figure 2 also shows the real position corresponding to the case $|x_0| = |y_0|$ with a black point and two segments, drawn with a black line, connecting this point to the points corresponding to the cases of $m_{real} = m* = |x_0|\ or|y_0|$. In Figure 2 the two segments are always under the computed magnitude arc. It should be noted that for all the input points belonging to the two segments, the corresponding magnitude m^*, estimated with Equation (3) and the circuit in Figure 1, is the radius of the light grey circumference arc in Figure 2.

$$m* = a\max(|x_0|, |y_0|) + b\min(|x_0|, |y_0|) \tag{3}$$

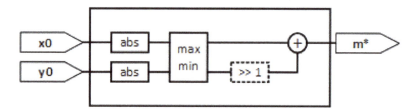

Figure 1. The fast magnitude estimator architecture.

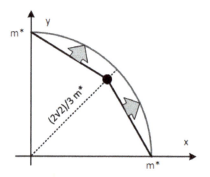

Figure 2. Segments for a computed magnitude of m^*.

If the x and y data paths are quantized on the M bit signed, the magnitude estimator in Figure 1 needs M bit unsigned to represent m^*. The idea is to segment the input space, based on m^*. We label as M1 area the set of points with the MSB (Most Significant Bit) of m^*, say $m^*[M-1]$, equal to one. If a point is not in the M1 area then we look for the $m^*[M-2]$ bit to mark it in the M2 area, and so on, until the desired level of segmentation is reached. Figure 3 shows the first area segments of the input coordinates while Figure 4 shows a simple circuit able to extract the logical information of the segmented area as a function of m^*.

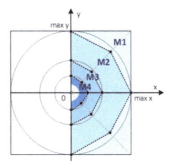

Figure 3. Segmentation of the input coordinates.

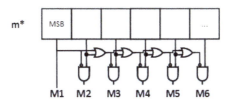

Figure 4. Segmentation flags circuit.

When the Mk area for a point is known, a shift factor (SFk) is selected and applied to the x_0 and y_0 input coordinates in a way that the magnified coordinates reach the M1 area or the M2 area. In other words, we want to work with points that are always in the first two segments, far from the zero point. Equation (4) shows the doubling rule to correctly shift the inputs starting from the M3 area (SF1 = 0, SF2 = 0). Figure 5 shows the architecture of the improved atan-CORDIC with the fast magnitude estimator and the input magnification. A difference compared to other CORDIC optimizations and CORDIC architectures in the literature [7–9,14–16] is the maintenance of the standard CORDIC core, to which we add a low-complexity pre-processing unit, working on the input ranges, thus minimizing the overall circuit complexity overhead.

$$x_0 2^{SF_k}; y_0 2^{SF_k} : SF_k = k - 2 \quad k \geq 3 \tag{4}$$

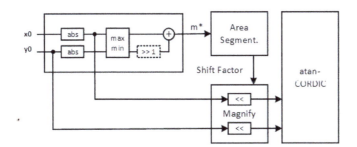

Figure 5. Improved atan-CORDIC architecture.

3. Computation Accuracy Evaluation

The architecture proposed in Figure 5 has been designed as a parametric HDL macrocell where the bit-width of the *x*, *y* and angle data paths can be customized at synthesis time, as well as the number n of CORDIC iterations and the number of input segments in Figure 3. For the atan-CORDIC core, a pipelined fully sequential architecture has been implemented. The choice of these parameters implies different optimization levels. An empirical parameter exploration is mandatory to characterize the sub-optimum architecture depending on the application requirements.

The quantization parameters in the fixed-point implementation are the number of bits of the x and y coordinates, the number of bits of the angle quantization, the number of iterations of the algorithm and the number of segmented areas. Giving the angle resolution in K bits, we know that the useful number of iterations comes from the relation where the quantized angle on the ATR set goes under half of its LSB (Last Significant Bit), being $LSB = \pi/2^{K-1}$. Overestimating the angle α_i with 2^{-i}, we can compute the useful limit of iterations with

$$2^{-i} < LSB/2 \Rightarrow i > K - \log_2 \pi \approx K - 1.65 \tag{5}$$

This means that the iteration index goes from 0 to $K - 1$ (K iterations). As a case study, we present an atan-CORDIC with an x, y data path on 14-bit signed, an angle data path on 12-bit signed and $n = 12$ iterations (iteration index from 0 to 11). To understand the phase error behavior, a first analysis is done holding the x coordinate ($x = 16$ in the example) and sweeping the entire y space. Figure 6 shows the absolute error in radians of the computed phase with the standard CORDIC with a blue trace. We see that for points near zero, the phase error increases significantly. The max error is ~0.10865 rad (about 71 LSB). Using the improved atan-CORDIC with nine area segments (max pre-magnification of 128), we get the result in Figure 7. In Figure 7, the blue trace is the absolute error. In Figure 7, the max error is ~0.0035 rad (less than 3 LSB). The same analysis can be extended through the x coordinates, collecting the maximum error of each run. Figure 8 shows the standard atan-CORDIC phase errors with a blue trace where the maximum is ~0.6355 rad (about 414 LSB). Figure 9 shows the improved atan-CORDIC errors with a blue trace where the maximum is ~0.0052 rad (less than 4 LSB). The red line in Figures 6–9 is the coarse upper-bound error evaluated with Equation (2).

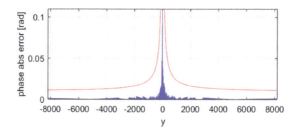

Figure 6. Standard atan-CORDIC phase error of $x = 16$ and y. The red line is the coarse upper-bound error evaluated with Equation (2); the blue trace is the standard atan-CORDIC phase error.

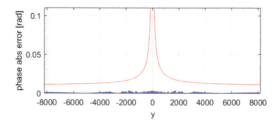

Figure 7. Improved atan-CORDIC phase error of $x = 16$ and y. The red line is the coarse upper-bound error evaluated with Equation (2); the blue trace is the standard atan-CORDIC phase error.

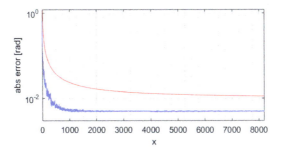

Figure 8. Standard atan-CORDIC max phase error on all the input space. The red line is the coarse upper-bound error evaluated with Equation (2); the blue trace is the standard atan-CORDIC phase error.

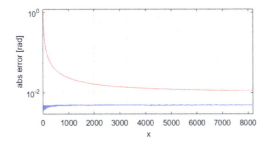

Figure 9. Improved atan-CORDIC max phase error on all the input space. The red line is the coarse upper-bound error evaluated with Equation (2); the blue trace is the standard atan-CORDIC phase error.

To evaluate further the performance of the proposed architecture, the maximum error on the angle is analyzed by varying some quantization parameters, as reported in Tables 1 and 2. In Table 1 the light cells show the standard CORDIC maximum error, fixed to 414 LSB (410 LSB in Table 2), while the grey cells represent the improved CORDIC maximum angle error. From the analysis of Tables 1 and 2, it is clear that increasing the number of M areas and/or increasing the bit width of the x and y data paths allows reducing the error. Instead, if we compare the peak error as a function of the number of iterations n (12 iterations in Table 1 vs. nine iterations in Table 2), the minimum is obtained for Table 2, which uses a number of iterations lower than the maximum 12 estimated from Equation (5), with $K = 12$. This can be justified by the fact that, as discussed in Section 1, the quantization in the data paths of the fixed-point implementation causes a truncation error that is accumulated in every loop of the algorithm.

Table 1. Max angle error expressed in LSB of the standard (std) vs. the proposed (our) CORDIC, $K = 12$, 12 iterations from 0 to 11, bits for x and y from 10 to 14, M areas from 6 to 9.

	M Areas							
	6		7		8		9	
x, y **Bits**	std	our	std	our	std	our	std	our
10	414	50	414	16	414	16	414	16
11	414	50	414	13	414	10	414	10
12	414	50	414	13	414	10	414	6
13	414	50	414	13	414	10	414	4
14	414	50	414	13	414	10	414	4

Table 2. Max angle error expressed in LSB of the standard (std) vs. the proposed (our) CORDIC, $K = 12$, nine iterations from 0 to 8, bits for x and y from 10 to 14, M-areas from 6 to 9.

	M areas							
	6		**7**		**8**		**9**	
x, y **Bits**	**std**	**our**	**std**	**our**	**std**	**our**	**std**	**our**
10	410	45	410	15	410	15	410	15
11	410	45	410	10	410	10	410	10
12	410	45	410	9	410	9	410	8
13	410	45	410	9	410	9	410	7
14	410	45	410	9	410	9	410	7

Indeed, in Table 3 we have reported, for both the standard and the proposed CORDIC, the maximum angle error, expressed in LSB, as a function of the number of iterations. For example, the minimum of the maximum angle error is achieved for the proposed CORDIC with seven iterations. This behavior can be theoretically justified, since the maximum angle error is achieved for low magnitude values, i.e., low values of the radius r in Equation (2). Figure 10 plots the error upper bound as a function of the number of iterations, using Equation (2) with $\epsilon = 1$, $\delta = \pi/2^{K-1}$, $K = 12$, and values for the radius *r* of 8, 16, 32 *and* 64. In Figure 10, similarly to what is observed in Table 3, there is a first part where the upper-bound error decreases and a second part where the upper-bound error increases.

Table 3. Max angle error expressed in LSB of the standard vs. proposed CORDIC ($K = 12$, 12 bits for *x* and *y* data paths, six M areas).

	Max. Number of Iterations										
	2	**3**	**4**	**5**	**6**	**7**	**8**	**9**	**10**	**11**	**12**
Standard CORDIC	512	302	252	333	374	394	404	409	412	413	414
Proposed CORDIC	512	302	160	82	43	30	40	45	48	49	50

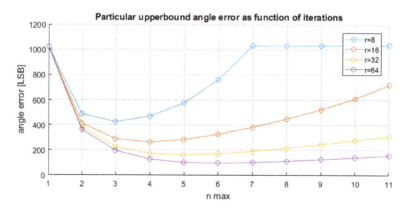

Figure 10. Upper-bound error behavior as function of the iteration number.

4. VLSI Implementation and Characterization

The improved CORDIC architecture proposed in Section 2 has been designed in VHDL as a parametric macrocell that can be integrated in any system-on-chip. With reference to the macrocell configuration discussed in Section 3, a synthesis in the standard-cell 180 nm 1.8 V CMOS technology from AMS ag has been carried out. The macrocell design is characterized by a low circuit complexity and low power consumption: about 1700 equivalent logic gates, with a max power consumption (static and dynamic power) of about 0.45 mW when working with a clock frequency of 30 MHz. The overhead due to the innovative parts of this work (fast magnitude estimation and input coordinates magnification) vs. a standard CORDIC implementation in the same technology is limited to 24% in terms of equivalent logic gates and −36% in terms of max working frequency. The processing time of the atan function is $n \cdot$Tclock, i.e., 0.4 μs when $n = 12$ with a clock of 30 MHz. As a possible improvement, a pipeline register can be inserted in Figure 5 before the atan-CORDIC core. This way, the complexity increases by about 10%, resulting in 1900 equivalent logic gates, while the working frequency can be increased up to 46 MHz and the power consumption is 0.77 mW. The processing time is $(n + 1) \cdot$Tclock; with $n = 12$ iterations of the atan function, the processing time is 0.28 μs.

The comparison of the proposed solution vs. the standard state-of-the-art solution, implemented in the same 180 nm 1.8 V technology, is summarized in Table 4. Table 4 shows also the comparison with other known CORDIC computation architectures [6,8]. In [8], an embedded solution is proposed using a 32-bit ART7TDMI-S core (18,000 equivalent gates) plus 32 kB of RAM (Random Access Memory) and 512 kB of EEPROM (Electrically Erasable Programmable Read Only Memory). With a clock of 60 MHz, the solution in [8] implements an atan evaluation in 35 μs and an angle error of 2.42×10^{-5} rad with an optimized algorithm. At 60 MHz with 180 nm 1.8 V CMOS technology, the ARM7-TDMIs has a power consumption of about 24 mW. Although [8] is optimized to achieve a lower error, 2.42×10^{-5} rad vs. 6.14×10^3 rad in our implementation, with $K = 12$, the proposed design outperforms the design in [8] in terms of the reduced processing time, by a factor $\times 124$, the reduced power consumption, by a factor $\times 31$, and the reduced circuit complexity, by a factor $\times 9.5$. To achieve a similar calculation accuracy as in [8], our IP has been re-synthetized with $K = 20$. In such a case, the processing time is still within 1 μs and the complexity in the logic gates is still below 4 kgates.

Table 4. Performance of the proposed CORDIC solution vs. standard solution.

IP macrocell	Standard CORDIC	This work	[6]	[8]
Equivalent gates	1400	1700 (1900 pipelined)	8000	18,000 + 32kB RAM + 512 kB E2PROM
Clock frequency, MHz	46	30 (46 pipelined)	10	60
Max. angle error	0.6355 rad	6.1×10^{-3} rad	N/A ($K = 16$, $n = 20$)	2.42×10^{-5} rad
Processing latency, μs	0.28	0.4 (0.28 pipelined)	N/A	35
Power consumption, mW	0.57	0.45 (0.77 pipelined)	N/A	24

The 180 nm technology node and the target clock frequency have been selected following the constraints of automotive applications where, to face a harsh operating environment (over-voltage, over-current, temperature up to 150 °C), a 180 nm 1.8 V technology ensures higher robustness vs. scaled sub-100 nm CMOS technologies operating with a voltage supply of 1 V. The considered 180 nm technology is automotive-qualified and has a logic gate density of 118 kgates/mm^2 [18]. The N-MOS supplied at 1.8 V has a voltage threshold of 0.35 mV and a current capability of 600 μA/μm^2. The P-MOS supplied at 1.8 V has a voltage threshold of 0.42 mV and a current capability of 260 μA/μm^2.

The proposed improved CORDIC IP has been integrated in an application-specific processor, called SENSASIP [5], to improve the digital signal processing capability of a 32-bit CORTEX-M0 core. Particularly, the SENSASIP chip is used to implement the algorithms in [3,19–21] in real time for the sensor signal processing of a 3D Hall sensor and of an inductive position sensor in X-by-wire automotive applications. In these specific applications, the real-time application requirements were met already with a clock frequency for SENSASIP of 8 MHz. Hence, with a 46 MHz clock frequency, the improved CORDIC architecture is about six times faster than required. At 8 MHz, the power consumption of the proposed CORDIC architecture is below 0.1 mW.

It should be noted that the CORDIC computation can be done via software on the 32-bit Cortex core, as is done in [15] for the same class of magnetic sensor position signal conditioning. However, as proved in [5], the use of the proposed CORDIC macrocell integrated on-chip allows for savings in power consumption up to 75%. It should be noted that the proposed macrocell can be also integrated in FPGA technology, as in [16], in case the application of interest requires an FPGA implementation rather than a system-on-chip one.

5. Conclusions

In this work, we propose an improved architecture for arctangent computation using the CORDIC algorithm. A fast magnitude estimator works with a pre-processing magnification to rescale the input coordinates to an optimum range as a function of the estimated magnitude. After the scaling, the inputs are processed by the standard CORDIC algorithm implemented without changing any data path precision. The improved architecture is generic and can be applied to any atan-CORDIC implementation. The architecture improvement is achieved with a low-circuit complexity since it does not require any special operation, but only comparison, shift and add operations. Several case studies have been analyzed, considering a resolution for x and y from 10 to 14 bits, a resolution of 12 bits for the angle, a number of iterations from two to 12, and from six to nine area segments. The bit-true analysis shows that the maximum absolute error in the atan-CORDIC computation improves by at least two orders of magnitude, e.g., in Table 1 from 414 LSB to 4 LSB. The proposed macrocell has been integrated in a system-on-chip, called SENSASIP, for automotive sensor signal processing. When implemented in an automotive-qualified 180 nm 1.8 V CMOS technology, two configurations are available. The first one has a circuit complexity of 1700 logic gates, a max clock frequency of 30 MHz, a CORDIC processing time of 0.4 µs and a power consumption of 0.45 mW. The second configuration has an extra pipeline stage, and has a circuit complexity of 1900 logic gates, a max clock frequency of 46 MHz, a CORDIC processing time of 0.28 µs and a power consumption of 0.77 mW. These values are compared well to state-of-the-art works in the same 180 nm technology node. The 46 MHz clock value is six times higher than what is required (8 MHz) by real-time automotive sensor signal conditioning applications. In these operating conditions, the power consumption of the proposed solution is below 0.1 mW.

Acknowledgments: This project was partially supported by the projects FP7 EU Athenis3D and the University of Pisa PRA E-TEAM.

Author Contributions: All authors contributed to the research activity and to the manuscript writing and review.

Conflicts of Interest: The authors declare no conflict of interest.

References

1. Volder, J.E. The CORDIC Trigonometric Computing Technique. *IRE Trans. Electr. Comput.* **1959**, *EC-8*, 330–334. [CrossRef]
2. Meher, P.K.; Valls, J.; Juang, T.B.; Sridharan, K.; Maharatna, K. 50 Years of CORDIC: Algorithms, Architectures, and Applications. *IEEE Trans. Circuit Syst. I* **2009**, *56*, 1893–1907. [CrossRef]
3. Sisto, A.; Pilato, L.; Serventi, R.; Fanucci, L. A design platform for flexible programmable DSP for automotive sensor conditioning. In Proceedings of the Nordic Circuits and Systems Conference (NORCAS): NORCHIP & International Symposium on System-on-Chip (SoC), Oslo, Norway, 26–28 October 2015; pp. 1–4.
4. Zheng, D.; Zhang, S.; Zhang, Y.; Fan, C. Application of CORDIC in capacitive rotary encoder signal demodulation. In Proceedings of the 2012 8th IEEE International Symposium on Instrumentation and Control Technology (ISICT), London, UK, 11–13 July 2012; pp. 61–65.
5. Sisto, A.; Pilato, L.; Serventi, R.; Saponara, S.; Fanucci, L. Application specific instruction set processor for sensor conditioning in automotive applications. *Microprocess. Microsyst.* **2016**, *47*, 375–384. [CrossRef]
6. Timmermann, D.; Hahn, H.; Hosticka, B.; Schmidt, G. A Programmable CORDIC Chip for Digital Signal Processing Applications. *IEEE J. Solid-State Circuits* **1991**, *26*, 1317–1321. [CrossRef]
7. Rajan, S.; Wang, S.; Inkol, R.; Joyal, A. Efficient approximation for the arctangent function. *IEEE Signal Process. Mag.* **2006**, *23*, 108–111. [CrossRef]

8. Ukil, A.; Shah, V.H.; Deck, B. Fast computation of arctangent functions for embedded applications: A comparative analysis. In Proceedings of the 2011 IEEE International Symposium on Industrial Electronics (ISIE), Gdansk, Poland, 27–30 June 2011; pp. 1206–1211.
9. Revathi, P.; Rao, M.V.N.; Locharla, G.R. Architecture design and FPGA implementation of CORDIC algorithm for fingerprint recognition applications. *Procedia Technol.* **2012**, *6*, 371–378. [CrossRef]
10. Walther, J.S. A unified algorithm for elementary functions. In Proceedings of the Spring Joint Computer Conference AFIPS '71 (Spring), Atlantic City, NJ, USA, 18–20 May 1971; pp. 379–385.
11. Hu, Y.H. The quantization effects of the CORDIC algorithm. *IEEE Trans. Signal Process.* **1992**, *40*, 834–844. [CrossRef]
12. Kota, K.; Cavallaro, J.R. A normalization scheme to reduce numerical errors in inverse tangent computations on a fixed-point CORDIC processor. In Proceedings of the 1992 IEEE International Symposium on Circuits and Systems, San Diego, CA, USA, 10–13 May 1992; Volume 1, pp. 244–247.
13. Fanucci, L.; Rovini, M. A Low-Complexity and High-Resolution Algorithm for the Magnitude Approximation of Complex Numbers. *IEICE Trans. Fundam. Electron. Commun. Comput. Sci.* **2002**, *E85-A*, 1766–1769.
14. Muller, J.-M. The CORDIC Algorithm. In *Elementary Functions*; Birkhäuser Boston: Cambridge, MA, USA, 2016; pp. 165–184.
15. Glascott-Jones, A. Optimising Efficiency using Hardware Co-Processing for the CORDIC Algorithm. In *Advanced Microsystems for Automotive Applications*; Springer: Heidelberg, Germany, 2010; pp. 325–335.
16. Taylor, A.P. How to use the CORDIC algorithm in your FPGA design. *XCell J.* **2012**, *79*, 50–55.
17. Anoop, C.V.; Betta, C. Comparative Study of Fixed-Point and Floating-Point Code for a Fixed-Point Micro. In Proceedings of the dSPACE User Conference 2012, Bangalore, India, 14 September 2012.
18. Minixhofer, R.; Feilchenfeld, N.; Knaipp, M.; Röhrer, G.; Park, J.M.; Zierak, M.; Enichlmair, H.; Levy, M.; Loeffler, B.; Hershberger, D.; et al. A 120 V 180 nm High Voltage CMOS smart power technology for System-on-chip integration. In Proceedings of the 2010 22nd International Symposium on Power Semiconductor Devices & IC's (ISPSD), Hiroshima, Japan, 6–10 June 2010.
19. Bretschneider, J.; Wilde, A.; Schneider, P.; Hohe, H.-P.; Koehler, U. Design of multidimensional magnetic position sensor systems based on hallinone technology. *IEEE Int. Symp. Ind. Electron. (ISIE)* **2010**, 422–427. [CrossRef]
20. Sarti, L.; Sisto, A.; Pilato, L.; di Piro, L.; Fanucci, L. Platform based design of 3D Hall sensor conditioning electronics. In Proceedings of the 2015 11th Conference on IEEE Conference on Ph.D. Research in Micr. and Electronics (PRIME), Glasgow, UK, 29 June–2 July 2015; pp. 200–203.
21. Kim, S.; Abbasizadeh, H.; Ali, I.; Kim, H.; Cho, S.; Pu, Y.; Yoo, S.-S.; Lee, M.; Hwang, K.C.; Yang, Y.; et al. An Inductive 2-D Position Detection IC With 99.8% Accuracy for Automotive EMR Gear Control System. *IEEE Trans. Very Large Scale Integr. (VLSI) Syst.* **2017**. [CrossRef]

Article

Automated Scalable Address Generation Patterns for 2-Dimensional Folding Schemes in Radix-2 FFT Implementations

Felipe Minotta* [ID]**, Manuel Jimenez** [ID] **and Domingo Rodriguez** [ID]

Electrical and Computer Engineering Department, University of Puerto Rico at Mayagüez,
Mayagüez 00681-9000, Puerto Rico; manuel.jimenez1@upr.edu (M.J.); domingo.rodriguez@upr.edu (D.R.)
* Correspondence: felipe.minotta@upr.edu

Received: 30 December 2017; Accepted: 1 March 2018; Published: 3 March 2018

Abstract: Hardware-based implementations of the Fast Fourier Transform (FFT) are highly regarded as they provide improved performance characteristics with respect to software-based sequential solutions. Due to the high number of operations involved in calculations, most hardware-based FFT approaches completely or partially fold their structure to achieve an efficient use of resources. A folding operation requires a permutation block, which is typically implemented using either permutation logic or address generation. Addressing schemes offer resource-efficient advantages when compared to permutation logic. We propose a systematic and scalable procedure for generating permutation-based address patterns for any power-of-2 transform size algorithm and any folding factor in FFT cores. To support this procedure, we develop a mathematical formulation based on Kronecker products algebra for address sequence generation and data flow pattern in FFT core computations, a well-defined procedure for scaling address generation schemes, and an improved approach in the overall automated generation of FFT cores. We have also performed an analysis and comparison of the proposed hardware design performance with respect to a similar strategy reported in the recent literature in terms of clock latency, performance, and hardware resources. Evaluations were carried on a Xilinx Virtex-7 FPGA (Field Programmable Gate Array) used as implementation target.

Keywords: Discrete Fourier Transform; Fast Fourier Transform; Linear Transform; Pease Factorization; scalable address generation; Digital Signal Processing; hardware generation

---

1. Introduction

The Fast Fourier Transform (FFT) is the main block in many electronic communications, signal processing, and scientific computing operations. It allows for the fast computation of the Discrete Fourier Transform (DFT). The DFT, in turn, is used to obtain the spectrum of any finite discrete signal or numeric sequence. Many FFT factorizations have been proposed since its first formulation, including Cooley-Tucky, Pease, and Hartley, among others, each with different advantages and limitations [1]. As regularity becomes a desirable characteristic when FPGA-based solutions are pursued, this work uses the Pease FFT formulation due to its *constant geometry* structure [2,3]. Our work formulates a set of algorithms for *addressing schemes* or *scheduling operations* in FFT computations. Figure 1a shows the standard representation of a 2-point FFT calculation, also called a *butterfly* operation. In Figure 1b we introduce an alternate butterfly notation which we will use for the butterfly operation. This notation facilitates the formulation of FFT algorithms in terms of Kronecker products algebra, introduced later in this paper. An N-point FFT operation is divided into S stages, where $S = \log_2(N)$ is a positive integer.

Figure 2 shows the structure of an 8-point Pease FFT implementation, illustrating a representative *N*-point structure where butterflies are presented as *processing elements* (PEs) in the rectangular boxes. Figure 2 also allows for introducing the concepts of horizontal and scalable vertical folding. A horizontal folding consists on reducing by half the number of columns in a FFT structure. A horizontal folding is considered maximum when successively halving the column count, reduces the total number of columns to one, maximizing the usability of hardware resources. The vertical folding consists on reducing the number of butterflies in the single column resulting resulting after a maximum horizontal folding has been applied. A vertical folding is considered scalable when the factor introduced to reduce the number of butterflies in the single column can be conveniently adjusted to fit a particular hardware resource limitation. A scalable vertical folding process affects the overall latency associated with memory operations. It also creates the need for a permutation block which is in charge of controlling the data flow between stages. A *vertical folding factor* ϕ is defined such that $\phi B = N$, where $B = 2\beta$ is defined as the number of memory banks required in the implementation, and β is defined as the number of processing elements in a single column. A permutation block consists of two data switches and B memory banks. For small values of β, the memory latency is high and the amount of consumed hardware resources is low. On the contrary, for large values of β, the latency is low and the amount consumed hardware resources is high. The resulting structure may be classified as a Burst I/O design since the computation must wait for the entire FFT calculation to finish before providing a new input signal to the hardware computational structure (HCS).

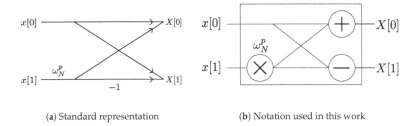

(a) Standard representation   (b) Notation used in this work

Figure 1. Representations for a 2-point Fast Fourier Transform (FFT) operation.

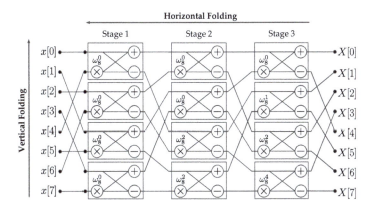

Figure 2. 8-point Pease FFT architecture.

Permutation blocks in folded hardware FFTs can be implemented in one of two ways: either using dedicated permutation logic or using address generation schemes. In the first approach, a complex dedicated logic circuit controls the data flow, while in the second, data are stored in memory banks and a simple controller generates the addresses depending on the needed permutations.

Several works have been reported using scalable permutation logic with variable number of points and folding factors. A major drawback of these approaches is the large amount of consumed hardware resources [4–6]. Addressing schemes were introduced by Johnson as an alternative able to significantly reduce the amount of consumed hardware resources [7]. Many FFT implementations have been reported to follow the concept of addressing schemes. Some have used a non-scalable approach targeted to specific applications [8–12]. Others provide scalable implementations in terms the number of FFT points [13–21].

Francheti et al. proposed a framework for the hardware generation of linear signal transforms [22]. Although their work produced functional solutions, this framework did not describe in sufficient detail the address generation scheme utilized in its table lookup procedure for twiddle factors or phase factors operations. It is essential to develop FFT computational algorithms with optimal data placement and access strategies, to obtain data accesses with zero conflicts. Data address generation schemes need to be designed to accomplish this task. Our address generation scheme satisfies this design requirement without the need of a reordering procedure, and it is independent of pipeline overlapping techniques, such as those reported in works by Richardson et al. [23].

Despite this diverse set of implementations, there are few approaches addressing the issue of using an HCS to allow for scalable folding processing. The main motivation behind this work is the existing need for a generalized addressing scheme for scalable algorithms based on maximum horizontal folding. To the best of our knowledge, no such generalized addressing scheme has been previously proposed. The key benefits of a our proposed addressing scheme can be summarized as follows:

- a set of methods to generate scalable FFT cores based on an address generation scheme for Field Programmable Gate Array (FPGA) implementation when the vertical folding factor is optimized,
- a mathematical procedure to automatically generate address patterns in FFT computations,
- a set of stride group permutations to guide data flow in hardware architectures, and
- a set of guidelines about space/time tradeoffs for FFT algorithm developers.

The rest of this paper is organized as follows: Section 2 describes the mathematical foundations of the FFT core. Section 3 describes the hardware implementation procedure. Section 4 evaluates the design performance in terms of clock latency and hardware resources when the target HCS is a Xilinx FPGA. Finally, Section 5 provides concluding remarks.

2. Mathematical Preliminaries

In this section, we start by presenting definitions and mathematical concepts used for the formulation of the proposed scalable address generation scheme.

2.1. Definitions

Let us first define two main matrix operations frequently used in this paper: Kronecker products and matrix direct sum. If $A \in \mathbb{R}^{m_1 \times n_1}$ and $B \in \mathbb{R}^{m_2 \times n_2}$, then their Kronecker product is defined as $C \in \mathbb{R}^{m_1 m_2 \times n_1 n_2}$ [24], where

$$C = A \otimes B = [a[k,l] \times B]_{k \in Z_{m_1}, l \in Z_{n_1}}. \tag{1}$$

The matrix direct sum of matrix set $A_1, A_2, ..., A_n$ is defined as

$$\bigoplus_{d=1}^{n} A_d = diag(A_1, A_2, ..., A_n) = \begin{bmatrix} A_1 & & & \\ & A_2 & & \\ & & \ddots & \\ & & & A_n \end{bmatrix}. \tag{2}$$

Now let us define the Pease formulation for an N-point FFT in Kronecker products,

$$F_N = \left\{ \prod_{i=1}^{S} (L_2^N (I_{N/2} \otimes F_2) T_i) \right\} R_N, \tag{3}$$

where L_2^N is a stride by 2 permutation of order N, $I_{N/2}$ is an identity matrix of size $N/2$, R_N is termed the bit reversal permutation of order N, and F_2 is the radix-2 butterfly defined as

$$F_2 = \begin{bmatrix} 1 & 1 \\ 1 & -1 \end{bmatrix}. \tag{4}$$

T_i is the twiddle or phase factor matrix defined as

$$T_i = \bigoplus_{r=0}^{2^{S-i}-1} (I_{2^{i-1}} \otimes W_2^q), \tag{5}$$

where $q = r \times 2^{i-1}$ and,

$$W_2^q = \begin{bmatrix} 1 & 0 \\ 0 & w_N \end{bmatrix}^q, \tag{6}$$

where $w_N = e^{-j\frac{2\pi}{N}}$ [25]. Furthermore, the direct computation shows that

$$R_N = (I_{N/2} \otimes L_2^2)(I_{N/4} \otimes L_2^4) \dots (I_2 \otimes L_2^{N/2}) L_2^N, \tag{7}$$

where L_2^N is, again, a stride by 2 permutation [26].

2.2. Kronecker Products Formulation of the Xilinx FFT Radix-2 Burst I/O Architecture

Our work is based on an equivalent architecture for the Xilinx FFT Radix-2 Burst I/O core (Figure 3) for implementing DFT computations [27]. Our goal is to present a systematic and scalable procedure for generating permutation-based address patterns for any power-of-2 transform size FFT algorithm and any folding factor. For this purpose, we propose to use the language of Kronecker products algebra as a mathematical tool. Therefore, we start by representing mathematically this architecture using a specific DFT computation example, for $N = 8$.

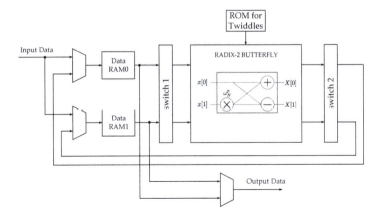

Figure 3. Xilinx Radix-2 Burst I/O architecture. RAM: random access memory; ROM: read only memory

At the beginning of the process, in the Xilinx architecture, data points are stored serially in RAM0 and RAM1 through the input data bus (Figure 3). The first 4 data points are stored in RAM0 and the last 4 are stored in RAM1. Following the Pease FFT formulation, the addressing schemes can be expressed as:

$$RA_0 = R_4 \begin{bmatrix} 0 & 1 & 2 & 3 \end{bmatrix}^T, \tag{8}$$

$$RA_{i+1} = L_2^4 RA_i \quad 0 \le i \le 2 , \tag{9}$$

$$RB_0 = R_4 \begin{bmatrix} 0 & 1 & 2 & 3 \end{bmatrix}^T, \tag{10}$$

$$RB_{i+1} = (J_2 \otimes I_2) L_2^4 RB_i \quad 0 \le i \le 2 , \tag{11}$$

where RA_i is the addressing scheme for RAM0, RB_i is the addressing scheme for RAM1, i is the *stage identifier*, and J_2 is a data exchange matrix. In the first stage, switch 1 in Figure 3 does not perform any permutation over the input data. In the rest of the stages, switch 1 does not perform any permutation in the first half, and interchanges the input data in the second half. That is,

$$
\begin{aligned}
SW1_{out} &= SW1_{in}; & i = 0, 0 \le b \le 3, \\
SW1_{out} &= SW1_{in}; & i > 0, 0 \le b \le 1, \\
SW1_{out} &= J_2 SW1_{in}; & i > 0, 1 \le b \le 3,
\end{aligned}
\tag{12}
$$

where $SW1_{out}$ is the output of switch 1, $SW1_{in}$ is the input of switch 1, b is the internal step counter. The single butterfly operation can be expressed as:

$$BF_{out} = \left[\left(\begin{bmatrix} 1 & 1 \\ 1 & 1 \end{bmatrix} \begin{bmatrix} 1 & 0 \\ 0 & w_8 \end{bmatrix}^p \odot \begin{bmatrix} 1 & 1 \\ 1 & -1 \end{bmatrix} \right) BF_{in} \right], \tag{13}$$

where BF_{out} and BF_{in} are the radix-2 butterfly's output and input in Figure 3, respectively. $w_8 = e^{-j\frac{2\pi}{8}}$ and,

$$p = 8 \times 2^{-(i+1)} \left\lfloor \frac{(2^{i+1}b)}{8} \right\rfloor, \tag{14}$$

where $\lfloor a \rfloor$ denotes the floor operation on a. Switch 2 in Figure 3 does not perform any permutation for the even-indexed values of b, the internal step counter. On the other hand, switch 2 interchanges the input data on the odd-values of b. That is,

$$
\begin{aligned}
SW2_{out} &= SW2_{in}; & i \ge 0, b = \{0, 2\}, \\
SW2_{out} &= J_2 SW2_{in}; & i \ge 0, b = \{1, 3\},
\end{aligned}
\tag{15}
$$

where $SW2_{out}$ is the output of the switch 2 and $SW2_{in}$ is the input of the switch 2. This process applies for any value of N. The equations included in this subsection are presented as an example of how a signal transform hardware architecture can be represented mathematically under a unified computational framework.

It is important to notice that this architecture only uses one butterfly. What would happen with the addressing schemes, switches, and phase factor scheduling if two or more butterflies were used? the next subsection addresses this relevant issue.

2.3. Kronecker Products Formulation of the Scalable FFT with Address Generation

Figure 4 shows an FFT architecture designed with the required blocks for our addressing scheme, under the control of a Finite State Machine (FSM). The *Data Address Generator* (DAG) produces the addresses depending on the needed permutations. The *Memory Banks* (MBs) store the input signal and the processed data as well. The *Data Switch Read* (DSR) permutates the data points from the MBs to process them correctly. The *Processing Elements* (PEs) perform the mathematical operations.

Finally, the *Data Switch Write* (DSW) permutates the data to avoid memory conflicts. Every block is mathematically represented as an entity, with specified inputs and outputs. To obtain these mathematical representations, we use again the language of matrix algebra and tensor or Kronecker products. In the next subsection, we define the general aspects of these mathematical representations.

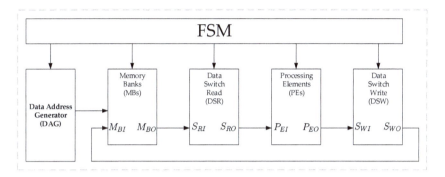

Figure 4. FFT architecture with our proposed addressing scheme. FSM: Finite State Machine.

The DAG must perform two kinds of permutations over the data points, regardless of the folding factor: a bit-reversal permutation in the first stage and stride-2 permutations in the next stages. Furthermore, since our implementation is intended to be used as a "conflict free memory addressing strategy" [7], these permutations have to be performed by rewriting the data points into the same locations as being read. Since each butterfly needs two different values at the same time, 2β different addressing schemes are needed. However, this work demonstrates that for any folding factor, only two array addressing schemes are needed. An *array addressing scheme* is defined as a systematic procedure for simultaneously pointing to B blocks of an ordered set of memory data. For an N-point FFT with vertical folding factor ϕ the array addressing schemes can be expressed as:

$$X_0 = R_\phi \begin{bmatrix} 0 & 1 & 2 & \dots & \phi - 1 \end{bmatrix}^T, \tag{16}$$

$$X_{i+1} = L_2^\phi X_i \quad 0 \leq i \leq S - 1, \tag{17}$$

$$Y_0 = R_\phi \begin{bmatrix} 0 & 1 & 2 & \dots & \phi - 1 \end{bmatrix}^T, \tag{18}$$

$$Y_{i+1} = (J_2 \otimes I_{\phi/2}) L_2^\phi Y_i \quad 0 \leq i \leq S - 1, \tag{19}$$

where X_i is the *even array addressing scheme*, Y_i is the *odd array addressing scheme*, i is the *stage identifier*, J_2 is an exchange matrix, L_2^ϕ and R_ϕ are, again, a stride permutation and a bit-reversal permutation, respectively, and the superscript T denotes matrix transpose operation.

The addressing scheme requires B memory banks. Since each processing element or butterfly accepts two complex numeric values at the same time, and there are β processing elements, the number of memory banks is equal to $B = 2\beta$. Each memory bank consists of ϕ memory locations. The memory banks can be indexed using the following scheme,

$$^b_i M_k[l]; k \in Z_B, l \in Z_\phi, \tag{20}$$

where $_i^b M_k$ is the k-th memory bank, and $0 \leq b \leq \phi - 1$ is the *internal step counter*. We use row major storage which means that each row contains the complex numeric values of an entire memory bank. That is,

$$
_i^b M = \begin{bmatrix}
M_0[0] & M_0[1] & \cdots & M_0[l] & \cdots & M_0[\phi-1] \\
M_1[0] & M_1[1] & \cdots & M_1[l] & \cdots & M_1[\phi-1] \\
\vdots & \vdots & \ddots & \vdots & \ddots & \vdots \\
M_k[0] & M_k[1] & \cdots & M_k[l] & \cdots & M_k[\phi-1] \\
\vdots & \vdots & \ddots & \vdots & \ddots & \vdots \\
M_{B-1}[0] & M_{B-1}[1] & \cdots & M_{B-1}[l] & \cdots & M_{B-1}[\phi-1]
\end{bmatrix}.
\tag{21}
$$

To extract the numeric values stored in the memory banks, we perform an inner product stage operation. This operation consists in first creating a *state indexing matrix*

$$
^i SIM_l[k]; l \in Z_\phi, k \in Z_B,
\tag{22}
$$

and then performing the *memory bank output* $(_b^i M_{BO})$ operation. We form this matrix by interleaving the even and odd array addressing schemes. That is,

$$
^i SIM = \underbrace{\begin{bmatrix} X_i & Y_i & X_i & \cdots & X_i & Y_i \end{bmatrix}}_{B},
\tag{23}
$$

where X_i and Y_i are, again, the even and odd array schemes vectors respectively. Furthermore, we use the notation $^i SIM_k[l]$ to refer the l-th element in k-th row in the i-th stage. Let E_ϕ be an ordered basis of ϕ column vectors with an entry of value one in the i-th position and zeros elsewhere. Let e_i, termed a *basis vector*, denote the i-th element of order ϕ, of this basis, where i ranges from 0 to $\phi - 1$. That is,

$$
E_\phi = \{e_0, e_1, ..., e_{\phi-1}\}.
\tag{24}
$$

The result of the inner product stage operation is the memory bank output $(_i^b M_{BO})$, and is obtained by making

$$
_i^b S_{RI} = {_i^b} M_{BO} = \begin{bmatrix}
\langle {_i^b} M_0, e_{i SIM_b[0]} \rangle \\
\langle {_i^b} M_1, e_{i SIM_b[1]} \rangle \\
\vdots \\
\langle {_i^b} M_k, e_{i SIM_b[k]} \rangle \\
\vdots \\
\langle {_i^b} M_{B-1}, e_{i SIM_b[B-1]} \rangle
\end{bmatrix},
\tag{25}
$$

where $_i^b S_{RI}$ is the DSR input bus, $e_{i SIM_b[k]} \in E_\phi$ is a basis vector, and

$$
_i^b M_k[^i SIM_b[k]] = \langle {_i^b} M_k, e_{i SIM_b[k]} \rangle = \sum_{n \in Z_\phi} {_i^b} M_k[n] (e_{i SIM_b[k]})[n].
\tag{26}
$$

The numeric values fetched throughout these addressing schemes are the ones to be operated by the PEs at each stage, but they are not in the same order according to the butterflies arrangement.

The DSR permutates these numeric values. In the first stage, the DSR performs a bit reversal permutation over the input data. In the rest of the stages, the DSR performs stride-2 permutations in the first half of the stage, and modified stride-2 permutations in the second. That is,

$$
\begin{aligned}
_i^b P_{EI} = {_i^b} S_{RO} = R_B {_i^b} S_{RI}; & \quad i = 0, 0 \leq b \leq \phi - 1, \\
_i^b P_{EI} = {_i^b} S_{RO} = L_{2 i}^{Bb} S_{RI}; & \quad i > 0, 0 \leq b \leq \phi/2 - 1, \\
_i^b P_{EI} = {_i^b} S_{RO} = (J_2 \otimes I_{\phi/2}) L_{2 i}^{Bb} S_{RI}; & \quad i > 0, \phi/2 \leq b \leq \phi - 1,
\end{aligned}
\tag{27}
$$

where S_{RO} is the DSR output bus, and P_{EI} is the PEs input bus. The next action is to process the DSR output through the butterflies. The PEs are an arrangement of radix-2 butterflies represented as F_2 and is expressed as:

$$
{}_i^b S_{WI} = {}_i^b P_{EO} = \begin{bmatrix} (U_2 W_2^p \odot F_2) \begin{bmatrix} {}_i^b P_{EI_0} \\ {}_i^b P_{EI_1} \end{bmatrix} \\ \\ (U_2 W_2^p \odot F_2) \begin{bmatrix} {}_i^b P_{EI_2} \\ {}_i^b P_{EI_3} \end{bmatrix} \\ \vdots \\ (U_2 W_2^p \odot F_2) \begin{bmatrix} {}_i^b P_{EI_k} \\ {}_i^b P_{EI_{k+1}} \end{bmatrix} \\ \vdots \\ (U_2 W_2^p \odot F_2) \begin{bmatrix} {}_i^b P_{EI_{B-2}} \\ {}_i^b P_{EI_{B-1}} \end{bmatrix} \end{bmatrix} , \tag{28}
$$

where ${}_i^b S_{WI}$ is the DSW input bus, U_2 is a unit matrix defined as

$$
U_2 = \begin{bmatrix} 1 & 1 \\ 1 & 1 \end{bmatrix}, \tag{29}
$$

and,

$$
W_2^p = \begin{bmatrix} 1 & 0 \\ 0 & \omega_N \end{bmatrix}^p, \tag{30}
$$

where $\omega_N = e^{-j\frac{2\pi}{N}}$ and,

$$
p = 2^{-(i+1)} N \left\lfloor \frac{(2^{i+1}\beta b + k)}{N} \right\rfloor . \tag{31}
$$

After the data points are processed by the butterflies, the results have to be stored in the same locations in the memory banks. However, to avoid memory bank conflicts, the results have to be permutated again. The DSW operator permutates these data points. For the even-indexed values of b, the internal step counter, the DSW does not perform any permutation. On the other hand, the DSW performs the next permutation on the odd-values of b. That is,

$$
\begin{aligned} {}_i^b M_{BI} = {}_i^b S_{WO} = {}_i^b S_{WI}; & \quad i \geq 0, b = 2n, n = \{0, 1, \ldots, \phi/2 - 1\}, \\ {}_i^b M_{BI} = {}_i^b S_{WO} = (I_{2\beta} \otimes J_2) {}_i^b S_{WI}; & \quad i \geq 0, b = 2n + 1, n = \{0, 1, \ldots, \phi/2 - 1\}, \end{aligned} \tag{32}
$$

where ${}_i^b S_{WO}$ is the DSW output bus, and ${}_i^b M_{BI}$ is the input bus of the memory banks.

Let G_ϕ be an ordered set of ϕ row vectors with a zero in the i-th position and ones elsewhere. Let g_i, termed a *masking vector*, denote the i-th element of order ϕ, of this set, where i ranges from 0 to $\phi - 1$. That is,

$$
G_\phi = \{g_0, g_1, \ldots, g_{\phi-1}\}. \tag{33}
$$

The next action is to write-back into the memory banks the numeric values contained in ${}_i^b S_{WO}$. This process can be done by making an addition. The operation is expressed as,

$$^{b+1}_i M = {}^b_i H \odot {}^b_i M + \begin{bmatrix} {}^b_i M_{BI_0} \left[e_{i SIM_b[0]} \right]^T \\ {}^b_i M_{BI_1} \left[e_{i SIM_b[1]} \right]^T \\ \vdots \\ {}^b_i M_{BI_k} \left[e_{i SIM_b[k]} \right]^T \\ \vdots \\ {}^b_i M_{BI_{B-1}} \left[e_{i SIM_b[B-1]} \right]^T \end{bmatrix}, \tag{34}$$

where the operator $\odot$ denotes Hadarmard or entrywise product, the superscript T means matrix transpose operation and,

$$^b_i H = \begin{bmatrix} g_{i SIM_b[0]} \\ g_{i SIM_b[1]} \\ \vdots \\ g_{i SIM_b[k]} \\ \vdots \\ g_{i SIM_b[B-1]} \end{bmatrix}, \tag{35}$$

where $g_{i SIM_b[k]} \in G_\phi$ is a masking vector. These formulations allow us to devise computational architectures associated with each expression. The next section describes how these expressions, called *functional primitives*, can be implemented as computational hardware architectures.

3. Hardware Implementation

3.1. Data Address Generator

The hardware implementation of this block consists mainly of a standard-order counter and two blocks which modify the counter values. One block generates the addressing scheme for the even-indexed memory banks (Equations (16) and (17)). The other generates the addressing scheme for the odd-indexed memory banks (Equations (18) and (19)). The even-indexed memory banks require stride-2 permutations calculated throughout the stages making a left-circular shift over the previous addressing scheme. An example of how to generate the entire addressing scheme is shown in Figure 5. Similarly, the odd-indexed memory banks require modified stride-2 permutations which are calculated throughout the stages by making a left-circular shift and negating the least significant bit over the previous addressing scheme. Figure 6 shows an example of how to generate the entire addressing schemes Figures 5 and 6 also show the decimal (N column) and decimal bit-reversal (R column) representations, respectively. Since it is easier to calculate the decimal representation, this is the one actually implemented. Nevertheless, the addressing scheme needed is the decimal bit-reversal representation. Therefore, all the address-generator outputs are bit-reversed to produce the correct addresses.

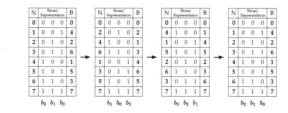

Figure 5. Addressing scheme example for even-indexed memories ($N = 16$ and $\beta = 1$).

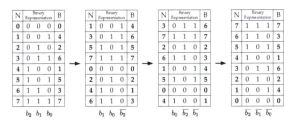

Figure 6. Addressing scheme example for odd-indexed memories ($N = 16$ and $\beta = 1$).

3.2. Phase Factor Scheduling

We took advantage of the periodicity and symmetry properties of the phase factors to reduce memory usage. Only $N/4$ numbers are stored, instead of $N/2$ as an FFT implementation normally requires [28]. The other $N/4$ values needed were obtained using,

$$\omega_N^{n+N/4} = \mathrm{Im}\{\omega_N^n\} - j\mathrm{Re}\{\omega_N^n\} \tag{36}$$

Tables 1 and 2 show the phase factor memory addresses. It is important to remark that the addressing scheme for the phase factors assumes memories of $N/2$ locations. Although we had established that the phase factor memories had $N/4$ locations. This is done due to the fact that the most significant bit of these addresses determines when Equation (36) is used. To establish an algorithm to generate the phase factors, we define three main parameters: the step (s), the frequency (f), and a parameter related with the initial value (n). The step s refers to the value to be added for obtaining the next address value, the frequency f refers to how often this step has to be added, and the initial value n is the first number in the addressing scheme. These main parameters behave as shown in Table 3. The frequency always starts at $N/(2\beta)$, decreases by halving its value in each iteration until the value is one, and remains as such until the calculation is over. The step always starts at $N/2$, and also decreases by halving its value in each iteration until the value reaches the number of butterflies β, and remains as such until the calculation is over. Finally, the parameter n, which represents the repetition and stride, at the same time, of the sequence of initial values, always stars at $N/4$, decreases by making a right shifting operation until the value is zero. Algorithm 1 generates the twiddles address pattern for an implementation of size N and β number of butterflies. In this Algorithm, T is the twiddle address, f is the frequency parameter, s is the step parameter, and n is the initial value.

Variable i represents the stage counter, b is the internal counter, and c is a counter. They are used to initialize the j-sequence per stage and to generate the addresses at each stage, respectively. Finally, R is a temporary variable used to store previous values of T.

Table 1. Phase factor addressing scheme example ($N = 32$ and $\beta = 1$).

Stage	Twiddle Address															
0	0	0	0	0	0	0	0	0	0	0	0	0	0	0	0	0
1	0	0	0	0	0	0	0	0	8	8	8	8	8	8	8	8
2	0	0	0	0	4	4	4	4	8	8	8	8	12	12	12	12
3	0	0	2	2	4	4	6	6	8	8	10	10	12	12	14	14
4	0	1	2	3	4	5	6	7	8	9	10	11	12	13	14	15

Table 2. Phase factor addressing scheme example ($N = 32$ and $\beta = 2$).

Stage	Twiddle Address							
0	0	0	0	0	0	0	0	0
	0	0	0	0	0	0	0	0
1	0	0	0	0	8	8	8	8
	0	0	0	0	8	8	8	8
2	0	0	4	4	8	8	12	12
	0	0	4	4	8	8	12	12
3	0	2	4	6	8	10	12	14
	0	2	4	6	8	10	12	14
4	0	2	4	6	8	10	12	14
	1	3	5	7	9	11	13	15

Table 3. Main parameters behaviour of the phase factor scheduling for $N = 32$.

a $\beta = 1$

Stage	s	f	n
0	16	16	8
1	8	8	4
2	4	4	2
3	2	2	1
4	1	1	0

b $\beta = 2$

Stage	s	f	n
0	16	8	8
1	8	4	4
2	4	2	2
3	2	1	1
4	1	1	0

c $\beta = 4$

Stage	s	f	n
0	16	4	8
1	8	2	4
2	4	1	2
3	4	1	1
4	4	1	0

Algorithm 1: Twiddle Address.

Input: $f = N/(2\beta)$, $s = N/2$, $n = N/2$

Output: Twiddle Pattern for N points and β butterflies

$T = 0$;

for $i = 1$ *to* $\log_2 N$ **do**

    $R = 0$;

    for $c = 0$ *to* $\beta - 1$ **do**

        $T = R$;

        if $c + 1$ *is divisible by* n *and* $T + n < N/2$ **then**

            $R = T + n$;

        end

        for $b = 2$ *to* $N/(2\beta)$ **do**

            if $b - 1$ *is divisible by* f **then**

                $T = T + s$;

            end

        end

    end

    $n = n/2$;

    if $f > 1$ **then**

        $f = f/2$;

    end

    if $s > \beta$ **then**

        $s = s/2$;

    end

end

3.3. Data Switch Read (DSR)

In hardware, the Data Switch Read (DSR) is implemented as an array of multiplexers, as shown in Figure 7. To perform the permutations, every multiplexer must choose the correct input. Equation (27) shows these permutations. The bit-reversal, stride-2, and modified stride-2 permutations are generated with bitwise operations over a standard-ordered sequence, such as circular shifting and simple logical operations. A bit-reversal permutation is obtained by making the *select input bus* (*SEL*) equal to

$$SEL = R_B \begin{bmatrix} 0 \\ 1 \\ \vdots \\ B-1 \end{bmatrix}, \tag{37}$$

where R_B is, again, a bit-reversal permutation.

A stride-2 permutation is obtained by making the *SEL* bus equals to

$$SEL = L_B^2 \begin{bmatrix} 0 \\ 1 \\ \vdots \\ B-1 \end{bmatrix}, \tag{38}$$

where L_B^2 is, again, a stride permutation. In hardware, this operation can be implemented by making a left circular shifting. A modified stride-2 permutation is also obtained by making the *SEL* bus equal to

$$SEL = (J_2 \otimes I_{B/2}) L_B^2 \begin{bmatrix} 0 \\ 1 \\ \vdots \\ B-1 \end{bmatrix}, \tag{39}$$

where J_2 and $I_{B/2}$ are, again, the exchange matrix and the identity matrix, respectively. In hardware, this operation can be implemented by making a left circular shifting only once, but negating the least significant bit.

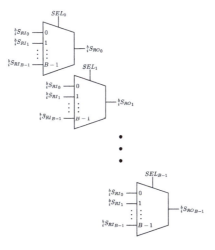

Figure 7. Hardware Implementation of the DSR.

3.4. Data Switch Write (DSW)

In hardware, the Data Switch Read (DSW) is also implemented as an array of multiplexers, as shown in Figure 8.

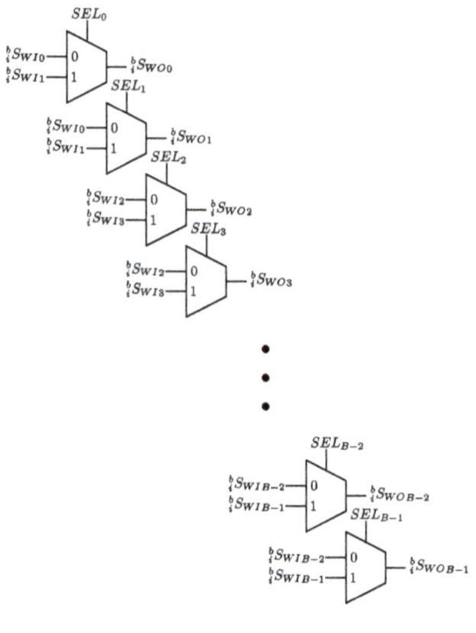

Figure 8. Hardware implementation of the DSW.

Equation (32) shows the required permutations. The first permutation outputs the input data in the same order. That is,

$$SEL = \begin{bmatrix} 0 \\ 1 \\ 0 \\ 1 \\ \vdots \\ 0 \\ 1 \end{bmatrix}. \tag{40}$$

The second permutation interchanges the even indexed input data with the next odd indexed input data. That is,

$$SEL = \begin{bmatrix} 1 \\ 0 \\ 1 \\ 0 \\ \vdots \\ 1 \\ 0 \end{bmatrix}. \tag{41}$$

3.5. Finite State Machine (FSM)

The FSM is a simple two-state Mealy machine. In first state, the machine is in idle state, waiting for an enable signal. The second state performs the address calculation, returning to the first state when $i = S - 1$ and $b = \phi - 1$. The FSM uses the values of i and b to successively generate the signals required to control:

- the output change of the DAG as indicated in Equations (17) and (19),
- the permutation to perform by the DSR and DSW as indicated in Equations (27) and (32),
- the phase factor addresses for the PEs and as indicated in Equation (28),
- the reading and writing process in the MBs as indicated in Equations (25) and (34).

4. FPGA Implementation

This section describes the procedure followed to validate the design. It shows the results obtained after completing the implementation of the proposed address generation scheme. The target hardware platform was a Xilinx VC-707 Evaluation Platform (Xilinx, Inc., San José, CA, United States). This particular board is based on a XC7VX485TFFG1761-2 FPGA (Xilinx, Inc., San José, CA, United States), which is a Virtex-7 family device. Resource consumption results were referenced for this specific chip. In order to provide a reference for comparison, a Xilinx FFT core v9.0 (Xilinx, Inc., San José, CA, United States) were included. Xilinx FFT IP core generator provides four architecture option: Pipelined Streaming I/O, Radix-4 Burst I/O, Radix-2 Burst I/O, and Radix-2 Lite Burst I/O. We concentrated on the Radix-2 architectures since they address the nature of our work. The Radix-2 Lite Burst I/O was not considered since is a simplified version of the Burst I/O version. Furthermore, the Pipelined Streaming I/O was not considered either since the architecture presented in this work optimizes the horizontal folding of any FFT computation.

It is important to emphasize that the Radix-2 Burst I/O architecture was designed to use a single butterfly performing simultaneously vertical and horizontal folding. Our architecture allows for a scalable vertical folding from $N/2$ butterflies (zero folding) to one butterfly (maximum folding). Hence, Radix-2 Burst I/O was chosen because it falls into the same category as the cores designed in this work. Our scalable vertical folding scheme may result in a time speed-up in a factor of β.

4.1. Validation

The implemented FFT with address generation scheme first went through a high level verification process. At this point, a MATLAB® program (MathWorks, Inc., Natick, MA, United States, 2014b) served to validate the correctness of the strategy by implementing the same dataflow that would be applied to the hardware version. After a hardware design was completed in VHSIC (very high speed integrated circuit) hardware description language VHDL,another program in MATLAB® was written with the purpose of generating random test data. These data were generated using the MATLAB® function *rand*, which generates uniformly distributed pseudorandom numbers. The data were then applied to the design by means of a text file. Then, the generated numbers were exported into MATLAB®, where they were converted into a suitable format for comparison with the reference FFT. After completing this procedure, we obtained the Mean Percent Error (*MPE*) of different implementations through the following formula:

$$MPE = 100 \times \frac{1}{N} \sum_{t=1}^{N} \frac{|a_t - h_t|}{h_t}, \tag{42}$$

where N is number of points, a_t is data generated by MATLAB®, and h_t is data generated by our FFT core in the target FPGA. This procedure was done for power of two transform sizes between 2^4 and 2^{14} points, using one-, two-, and four-butterfly versions, and using single precision floating point format. Figure 9 illustrates *MPE* of our core compared to the FFT calculated with MATLAB®. Figure 9 also shows that there is a minimum percent error of 1.5×10^{-5} on an FFT core of 16 points and a maximum percent error of 4.5×10^{-5} for an FFT core of 16,384 points. This information reflects an incrementing percent error while the number of points is increased. This is because MATLAB® uses double precision floating point format and rounding errors, which are produced by the large number of operations involved in the calculation. These results confirm that our scalable FFT core implemented using address generation scheme worked as expected.

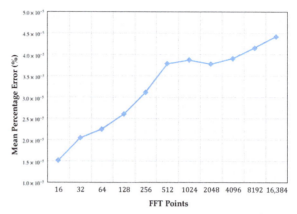

Figure 9. Mean percentage error of our core with respect to MATLAB®'s FFT function.

4.2. Timing Performance

In general, when applying the developed strategy, the number of clock cycles required to complete the operation is a well defined function of the transform size N and the number of butterflies used. This is, for a given transform of size N, with β butterflies, and a butterfly latency T_b, the cycles required to complete the calculation would be:

$$Cycles = \left(\frac{N}{2\beta} + T_b \right) \times (\log_2 N). \qquad (43)$$

Figure 10 shows the computation times comparison of the core implemented with the one-, two-, and four-butterfly versions and the Xilinx version running at 100 MHz as a function of the number of points.

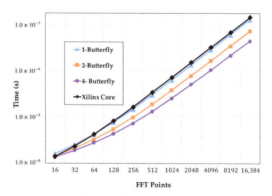

Figure 10. Computation time.

4.3. Resource Consumption

Figure 11 shows the number of slice registers consumed by the different implementations, starting with the single-butterfly to a four-butterfly version. Figure 12 shows the slice look up tables (LUTs) used. Finally, Figure 13 show the memory usage in terms of RAM blocks. The number of slices reflects the amount of logical resources spent. Flip-flops are contained within the slices and give information on how much resource is spent in sequential logic. The number of digital signal processing (DSP) blocks represents the amount of special arithmetic units from the FPGA dedicated to implement the butterfly. The total memory reflects the resources spent to store both, data points and phase factors. Regarding the DSP blocks consumed, each butterfly in this work uses 24 blocks, 50% more than the used by the Xilinx Core. The XC7VX485TFFG1761-2 has 607,200 slice registers, 303,600 Slice LUTs, and 2800 DSP blocks.

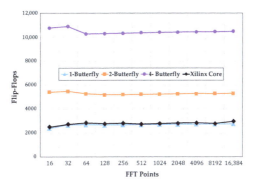

Figure 11. Flip-Flops consumption comparison.

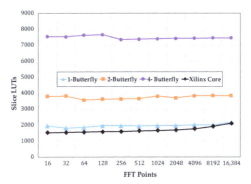

Figure 12. Slice look up tables (LUTs) consumption comparison.

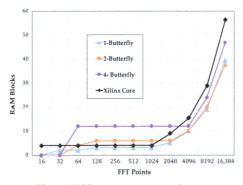

Figure 13. Memory usage comparison.

4.4. Analysis

Figure 13 reveals that the most heavily affected resource was the memory, as its usage increased in proportion to the transform size. However, using our addressing scheme did not affect the consumption

79

of memory resources. Figure 13 highlights this fact by showing that for increasing levels of parallelism the memory requirements were unaffected. Moreover, for a specific input size and phase factor, the data and address bus width increased proportionally to $\log_2(N)$; therefore, the system blocks managing these buses would not grow aggressively. The one-butterfly version had a smaller slice register consumption and almost the same slice count as the LUTs used by the FFT Xilinx Core. However, the two- and four-butterfly versions had a higher slice register and slice LUTs use. This was the trade-off to obtain a lower latency.

5. Conclusions

This paper proposed a method to generate scalable FFT cores based on an address generation scheme for FPGA implementation when the folding factor is scaled. The proposed method is an effective alternative to achieve a scalable folding of the Pease FFT structure. As a novel contribution, we provide a mathematical formulation to generate the address patterns and to describe data flow in FFT core computations.

Our approach is complete in the sense that it takes into consideration the address sequences required to access data points as well as twiddle or phase factors. This is accomplished by providing an algorithm and hardware to reproduce the twiddle factors for the Pease FFT radix-2 factorization regardless the number of points, folding factor, and numeric format using addressing schemes. Our implementation improves the computation time about 20% with one butterfly, 46% with two butterflies and 58% with four butterflies when compared to a reference design of the FFT Xilinx Core.

Acknowledgments: This work was supported in part by Texas Instruments Inc. through the TI-UPRM Collaborative Program. The views expressed in this manuscript are the opinions of the authors and do not represent any official position of any institution partially supporting this research.

Author Contributions: Felipe Minotta conceived, designed, and performed the experiments; Manuel Jimenez contributed to analyze the data; Domingo Rodriguez contributed with the mathematical formulations. All authors contributed to the writing of the paper.

Conflicts of Interest: The authors declare no conflict of interest.

References

1. Cooley, J.W.; Tukey, J. An Algorithm for the Machine Calculation of Complex Fourier Series. *Math. Comput.* **1965**, *19*, 297–301.
2. Pease, M.C. An Adaptation of the Fast Fourier Transform for Parallel Processing. *J. ACM* **1968**, *15*, 252–264.
3. Astola, J.; Akopian, D. Architecture-oriented regular algorithms for discrete sine and cosine transforms. *IEEE Trans. Signal Process.* **1999**, *47*, 1109–1124.
4. Chen, S.; Chen, J.; Wang, K.; Cao, W.; Wang, L. A Permutation Network for Configurable and Scalable FFT Processors. In Proceedings of the IEEE 9th International Conference on ASIC (ASICON), Xiamen, China, 25–28 October 2011; pp. 787–790.
5. Montaño, V.; Jimenez, M. Design and Implementation of a Scalable Floating-point FFT IP Core for Xilinx FPGAs. In Proceedings of the 53rd IEEE International Midwest Symposium on Circuits and Systems (MWSCAS), Seattle, WA, USA, 1–4 August 2010; pp. 533–536.
6. Yang, G.; Jung, Y. Scalable FFT Processor for MIMO-OFDM Based SDR Systems. In Proceedings of the 5th IEEE International Symposium on Wireless Pervasive Computing (ISWPC), Modena, Italy, 5–7 May 2010; pp. 517–521.
7. Johnson, L.G. Conflict Free Memory Addressing for Dedicated FFT Hardware. *IEEE Trans. Circuits Syst. II Analog Digit. Signal Process.* **1992**, *39*, 312–316.
8. Wang, B.; Zhang, Q.; Ao, T.; Huang, M. Design of Pipelined FFT Processor Based on FPGA. In Proceedings of the Second International Conference on Computer Modeling and Simulation, ICCMS '10, Hainan, China, 22–24 January 2010; Volume 4, pp. 432–435.

9. Polychronakis, N.; Reisis, D.; Tsilis, E.; Zokas, I. Conflict free, parallel memory access for radix-2 FFT processors. In Proceedings of the 19th IEEE International Conference on Electronics, Circuits and Systems (ICECS), Seville, Spain, 9–12 December 2012; pp. 973–976.

10. Huang, S.J.; Chen, S.G. A High-Throughput Radix-16 FFT Processor With Parallel and Normal Input/Output Ordering for IEEE 802.15.3c Systems. *IEEE Trans. Circuits Syst. I Regul. Papers* **2012**, *59*, 1752–1765.

11. Chen, J.; Hu, J.; Lee, S.; Sobelman, G.E. Hardware Efficient Mixed Radix-25/16/9 FFT for LTE Systems. *IEEE Trans. Large Scale Integr. (VLSI) Syst.* **2015**, *23*, 221–229.

12. Garrido, M.; Sanchez, M.A.; Lopez-Vallejo, M.L.; Grajal, J. A 4096-Point Radix-4 Memory-Based FFT Using DSP Slices. *IEEE Trans. Large Scale Integr. (VLSI) Syst.* **2017**, *25*, 375–379.

13. Xing, Q.J.; Ma, Z.G.; Xu, Y.K. A Novel Conflict-Free Parallel Memory Access Scheme for FFT Processors. *IEEE Trans. Circuits Syst. II Express Briefs* **2017**, *64*, 1347–1351.

14. Xia, K.F.; Wu, B.; Xiong, T.; Ye, T.C. A Memory-Based FFT Processor Design With Generalized Efficient Conflict-Free Address Schemes. *IEEE Trans. Large Scale Integr. (VLSI) Syst.* **2017**, *25*, 1919–1929.

15. Gautam, V.; Ray, K.; Haddow, P. Hardware efficient design of Variable Length FFT Processor. In Proceedings of the 2011 IEEE 14th International Symposium on Design and Diagnostics of Electronic Circuits Systems (DDECS), Cottbus, Germany, 13–15 April 2011; pp. 309–312.

16. Tsai, P.Y.; Lin, C.Y. A Generalized Conflict-Free Memory Addressing Scheme for Continuous-Flow Parallel-Processing FFT Processors With Rescheduling. *IEEE Trans. Large Scale Integr. (VLSI) Syst.* **2011**, *19*, 2290–2302.

17. Xiao, X.; Oruklu, E.; Saniie, J. An Efficient FFT Engine With Reduced Addressing Logic. *IEEE Trans. Circuits Syst. II Express Briefs* **2008**, *55*, 1149–1153.

18. Shome, S.; Ahesh, A.; Gupta, D.; Vadali, S. Architectural Design of a Highly Programmable Radix-2 FFT Processor with Efficient Addressing Logic. In Proceedings of the International Conference on Devices, Circuits and Systems (ICDCS), Coimbatore, India, 15–16 March 2012; pp. 516–521.

19. Ayinala, M.; Lao, Y.; Parhi, K. An In-Place FFT Architecture for Real-Valued Signals. *IEEE Trans. Circuits Syst. II Express Briefs* **2013**, *60*, 652–656.

20. Qian, Z.; Margala, M. A Novel Coefficient Address Generation Algorithm for Split-Radix FFT (Abstract Only). In Proceedings of the 2015 ACM/SIGDA International Symposium on Field-Programmable Gate Arrays, Monterey, California, USA, 22–24 February 2015; ACM: New York, NY, USA, 2015; p. 273.

21. Yang, C.; Chen, H.; Liu, S.; Ma, S. A New Memory Address Transformation for Continuous-Flow FFT Processors with SIMD Extension. In Proceedings of CCF National Conference on Compujter Engineering and Technology, Hefei, China, 18–20 October 2015; Xu, W., Xiao, L., Li, J., Zhang, C., Eds.; Springer: Berlin/Heidelberg, Germany, 2016; pp. 51–60.

22. Milder, P.; Franchetti, F.; Hoe, J.C.; Püschel, M. Computer Generation of Hardware for Linear Digital Signal Processing Transforms. *ACM Trans. Des. Autom. Electron. Syst.* **2012**, *17*, 15.

23. Richardson, S.; Marković, D.; Danowitz, A.; Brunhaver, J.; Horowitz, M. Building Conflict-Free FFT Schedules. *IEEE Trans. Circuits Syst. I Regul. Papers* **2015**, *62*, 1146–1155.

24. Loan, C.F.V. The ubiquitous Kronecker product. *J. Comput. Appl. Math.* **2000**, *123*, 85–100. Numerical Analysis 2000. Vol. III: Linear Algebra.

25. Johnson, J.; Johnson, R.; Rodriguez, D.; Tolimieri, R. A Methodology for Designing, Modifying, and Implementing Fourier Transform Algorithms on Various Architectures. *Circuits Syst. Signal Process* **1990**, *9*, 450–500.

26. Rodriguez, D.A. On Tensor Products Formulations of Additive Fast Fourier Transform Algorithms and Their Implementations. Ph.D. Thesis, City University of New York, New York, NY, USA, 1988.

27. Xilinx, Inc. https://www.xilinx.com/support/documentation/ip_documentation/xfft/v9_0/pg109-xfft.pdf (accessed on 20 October 2017).

28. Polo, A.; Jimenez, M.; Marquez, D.; Rodriguez, D. An Address Generator Approach to the Hardware Implementation of a Scalable Pease FFT Core. In Proceedings of the IEEE 55th International Midwest Symposium on Circuits and Systems (MWSCAS), Boise, ID, USA, 5–8 August 2012; pp. 832–835.

Sample Availability: Samples of the compounds ...... are available from the authors.

 *electronics*

Article

Exploiting Hardware Vulnerabilities to Attack Embedded System Devices: a Survey of Potent Microarchitectural Attacks

Apostolos P. Fournaris [1,2,*], Lidia Pocero Fraile [1] and Odysseas Koufopavlou [1]

[1] Electrical and Computer Engineering Department University of Patras, Rion Campus, Patras 26500, Greece; lidiapf@upatras.gr (L.P.F.); odysseas@ece.upatras.gr (O.K.)

[2] Industrial Systems Institute, Research Center ATHENA, Platani, Patra 26504, Greece

* Correspondence: apofour@ece.upatras.gr; Tel.: +30-261-099-6822

Received: 31 May 2017; Accepted: 4 July 2017; Published: 13 July 2017

Abstract: Cyber-Physical system devices nowadays constitute a mixture of Information Technology (IT) and Operational Technology (OT) systems that are meant to operate harmonically under a security critical framework. As security IT countermeasures are gradually been installed in many embedded system nodes, thus securing them from many well-know cyber attacks there is a lurking danger that is still overlooked. Apart from the software vulnerabilities that typical malicious programs use, there are some very interesting hardware vulnerabilities that can be exploited in order to mount devastating software or hardware attacks (typically undetected by software countermeasures) capable of fully compromising any embedded system device. Real-time microarchitecture attacks such as the cache side-channel attacks are such case but also the newly discovered Rowhammer fault injection attack that can be mounted even remotely to gain full access to a device DRAM (Dynamic Random Access Memory). Under the light of the above dangers that are focused on the device hardware structure, in this paper, an overview of this attack field is provided including attacks, threat directives and countermeasures. The goal of this paper is not to exhaustively overview attacks and countermeasures but rather to survey the various, possible, existing attack directions and highlight the security risks that they can pose to security critical embedded systems as well as indicate their strength on compromising the Quality of Service (QoS) such systems are designed to provide.

Keywords: embedded system security; microarchitecture attacks; cache attacks; rowhammer attack

---

1. Introduction

Embedded systems are gradually gaining a considerable market and technology share of computational system devices in various domains. Some of those domains, like Critical Infrastructure (CI) environments, have a need for fast responsiveness thus requiring real-time embedded system but in parallel, they also have a high demand for strong security. Cybersecurity attackers can find fertile ground in the embedded system domain since such systems' original structure was not meant to provide security but rather high QoS and in several occasions real time response. Under this framework, actual QoS in a real-time embedded system can only be achieved when security is included as a QoS parameter.

A wide range of attacks from the cyber security domain can be mounted on real-time embedded system devices. Many of them are thwarted by IT based countermeasures including antimalware programs, firewalls, Intrusion Detection Systems and Anomaly detection tools. Recently, in the H2020 European CIPSEC (Enhancing Critical Infrastructure Protection with Innovative SECurity Framework) project, the above tools are considered as a unified whole, thus structuring a cyberattack protection framework that is applicable to many different Critical infrastructure environments that involve real-time embedded systems [1]. In this framework, apart from Critical Infrastructure, end devices mostly being embedded systems are also identified as a possible target of a security attacker. Thus, security strengthening of embedded system end devices is suggested through both hardware and software means.

Embedded systemend nodes, as most computing units, cannot be considered trusted due to many known vulnerabilities. Typically, software tools (e.g., antivirus, antimalware, firewalls) can protect a system from attackers taking advantage of those vulnerabilities to inject some malicious software. However, there exist security gaps and vulnerabilities that cannot be traced by the above-mentioned software tools since they are not stationed on software applications, drivers or the Operating System but rather on the computer architecture and hardware structure itself. Creating software to exploit these vulnerabilities remains hidden from most software cybersecurity tools and thus constitute a serious security risk for all devices using these commodity, vulnerable, computer structures.

One of the most potent means of enhancing computer security without compromises in QoS (i.e., real-time responsiveness, computation and network performance, power consumption etc.) is to migrate security functions in hardware. For this reason, there exist many hardware security elements along with appropriate software libraries that advertise security and trust. Trusted Platform Modules creating Trusted Execution Environment are very strong trust establishment solutions that provide secure, isolated from potential attacks, execution of sensitive applications [2–4]. This is achieved by creating isolated memory areas to execute code and store sensitive data, as proposed and implemented by ARM processor TrustZone technology [4]. These solutions, that are gradually been introduced in security critical areas (e.g., in CI systems), however, cannot be fully protected against hardware-based vulnerabilities.

Typical computer hardware components, like memory modules and caches, exhibit strange behavior under specific circumstances thus enabling backdoor access to potential attackers. Memory disturbance errors are observed in commodity DRAM chips stationed and used in all CI devices. It was pointed out that when a specific row of a DDR (Double Data Rate) memory bank is accessed repeatedly, i.e., is opened (i.e., activated) and closed (i.e., precharged) within a DRAM refresh interval, one or more bit flips occur in physically-adjacent DRAM rows to a wrong value. This DRAM disturbance error has been known as RowHammer [5].

Several researchers have observed that Rowhammer can be exploited to mount an attack and bypass most of the established software security and trust features including memory isolation by writing appropriate malicious software thus introducing memory disturbance attacks or RowHammer attacks. Such attacks can be used to corrupt system memory, crash a system, obtain and modify secret data or take over the entire system.

Furthermore, the security related computation on a real-time embedded system device can leak the sensitive information being processed to a knowledgeable attacker. These types of attacks, denoted as side channel attacks (SCAs) aim at collecting leakage data during processing and through statistical computations extract sensitive information. SCAs can be very potent to embedded system devices left unwatched for long periods of time so that an attacker can have physical access or proximity to them. Characteristic examples of such unattended devices are critical infrastructure systems' end nodes (e.g., sensors, in field actuators, PLCs (PLC: Programmable Logic Controller) or other monitoring/control devices). However, even if physical access to devices is not possible, there exist SCA types, like microarchitectural/cache SCAs, that can be mounted remotely.

Under the light of the above dangers, focused on a device's architecture and hardware structure, embedded system devices that are traditionally designed and built mainly for high QoS (reliability, high response time, high performance) and real-time responsiveness are left unprotected. In this paper, an overview of this attack research field is provided focusing attack directions and countermeasures. We provide a description and analysis of the attack possibilities and targets that exploit the rowhammer vulnerability as well as microarchitectural/cache SCAs that can be mounted from software and use computer architecture leakage channels for exploitation. While the above attacks can also be mounted remotely, in this paper, for the sake of completeness, a small overview is also provided for other SCAs that are very powerful when physical access to a device under attack is possible. All the attacks in this paper can bypass traditional cybersecurity software tools. The goal of this work is to help real time embedded system security experts/architects understand the above-mentioned attacks and motivate them to take appropriate countermeasures as those sketched in the paper.

The rest of the paper is organized as follows. In Section 2, the Rowhammer vulnerability is defined and its triggering mechanism is described. In Section 3, attacks using Rowhammer vulnerability are described for x86 and ARM embedded systems. In Section 4 side channel attacks that are applicable to embedded system devices are described focusing on microarchitecture attacks that can be mounted remotely. Finally, Section 5 provides conclusions and future research directions in this research field.

2. The Rowhammer Vulnerability

In the quest to get memories smaller, to reduce the cost-per-bit, and make them faster, vendors have reduced the physical geometry of DRAMs and increased density on the chip. But smaller cells can hold a smaller, limited amount of charge which reduces its noise margin and renders it more vulnerable to data loss. Also, the higher cell's proximity introduces electromagnetic coupling effects between them. And the higher variation in process technology increases the number of cells susceptibility to inter-cell crosstalk. Therefore, new DRAM technologies are more likely to suffer from disturbance that can go beyond their merges and provoke errors.

The existence and widespread nature of disturbance errors in the actual DRAM chips were exposed firstly in [5] at 2014. In this work, the root of the problem was identified in the voltage fluctuation on internal wires called wordline. Each row has its own *wordline* whose voltage is rising on a row access. Then, many accesses to the same row force this line to be toggled on and off repeatedly provoking voltage fluctuation that induce a disturbance effect on nearby rows. The perturbed rows leak charge at accelerated rate and if its data is not restored fast enough some of the cells changes its original value. In fact, it has been found that this vulnerability exists in the majority of the recent commodity DRAM chips, being more prevalent on 40 nm memory technologies. In the work of Carnegie Mellon University [5] this error phenomenon was found to exist in 139 DDR3 DRAM modules on x86 processors, for all the manufacturer modules from 2013 to 2014. Lanteigne performed an analysis on DDR4 memory [6] proving that rowhammer are certainly reproducible on such technology too, in March 2016. From the 12 modules under test, 8 of them show bit flips during theirs experiments under default refresh rate. And most recently in the paper [7], where a determinist Rowhammer Attack has been proposed for Mobile Platforms, it is observed that the majority of LPDDR3-devices (Low Power DDR memory devices) under test have induced flips and that even LPDDR2 is vulnerable, under ARMv7 based processors. Also, DDR4 memory modules, that include several countermeasures, remain vulnerable to the Rowhammer bug, as described in [8] where a Rowhammer based attack on DDR4 vulnerability is proposed.

The reasons behind the rowhammer vulnerability bug can be traced back to the diverse ways of interaction that provoke this specific disturbance as hypothesized in [5]. Note that this is not based on DRAM chip analysis but rather on past studies. Changing the voltage of a wordline could inject noise into an adjacency wordline through electromagnetic coupling and partially enabling it can cause leakage on its row's cells.

Toggling the wordline could accelerate the flow of charge between two bridge cells. Bridges are a class of DRAM faults in which conductive channels are formed between unrelated wires or capacitors. The hot-carrier injection effect can permanently damage a wordline and can be produced by toggling for 100 hours the same wordline as proven in [5]. Then, some of the hot-carrier can be injected on a neighboring row and modify their cells charge or alter the characteristics of their access transistors.

Triggering Rowhammer

By repeatedly accessing, hammering, the same memory row (aggressor row) an attacker can cause enough disturbance in a neighboring row (victim row) to cause a bit flip. This can be software triggering by code with a loop that generates millions of reads to two different DRAM rows of the same bank in each iteration.

The ability to make this row activation fast enough is the first prerequisite to trigger the bug. The Memory Controller must allow this fast access but generally, the biggest challenge is overpassing the several layers of cache that mask out all the CPU (Central Processing Unit) memory reads.

A memory access consists of stages. In ACTIVE stage, firstly a row is activated so as to transfer the data row to the bank's row buffer by toggling ON its specific associated wordline. Secondly, the specific column from the row is read/written (READ/WRITE stage) from or to the row buffer. Finally, the row is closed, by precharging (PRECHARGE stage) the specific bank, writing back the value to the row and plugging OFF the wordline. The disturbance error is produced on a DRAM row when a nearby wordline voltage is toggled repeatedly, meaning that it is produced on the repeated ACTIVE/PRECHARGE of rows and not on the column READ/WRITE stage. When a software code is triggering the Rowhammer bug by repeated accesses to the same physical DRAM address, care must be taken in order to ensure that each access will correspond to a new row activation. If the same physical address is accessed continuously, the corresponding data is already in the row buffer and no new activation is produced. Therefore, we must access two physical addresses that correspond to rows on the same bank to be sure that the *row buffer* is cleaning between memory access.

A second challenge to consider is the need of finding physical addresses that are mappings on rows from the same bank (single-side rowhammer*)*. Early projects confronted this challenges by picked random addresses following probabilistic approaches [5,9–11]. In latest works, the exact physical address mapping of all the banks should be known in order to access both rows that are directly above and below the *victim* one to mount a *double-side rowhammer* attack. This attack is much more efficient and can guarantee a deterministic flip of a bit, on vulnerable DRAM memory, for specific chosen victim address location [7,8,12].

The capability to induce disturbance errors on specific systems depends on the toggle rate in relation with the refresh interval. This refers to how much times a specific row is activated between *refresh* commands that recharge the data on each cell. Generally, the more times the row is open on a refresh interval the bigger is the probability of bit flips. Nevertheless, when the memory access rate overpasses a limit, smaller number of bits are being flip than the typical case [7]. The memory controller could schedule refreshes and memory access commands under pressing memory accessing conditions [5]. Then, No Operation Instructions (NOP) on each integration of the loop can help sometimes to trigger more flip bits in order to lower the access memory rate to the same address [7]. On other hand, at a sufficiently high refresh interval, the errors could be completely eliminated but with a corresponding consequence on both power consumptions and performance [5,7,13].

Depending on the DRAM implementation and due to the intrinsic orientation property, some cells (true-cells) represent a logical value of '1' using the charge state while others cell (anti-cells) represent a logical value of '1' using the discharge state. There are cases where both orientations are used in distinct parts of the same DRAM module. For triggering the flip on bits, the cells must be charged to increase its discharge rate in order to trigger rowhammer. This means that only logical value '1' can produce flips for *true-cells* while logical value '0' data can produce flips for *anti-cells*. Therefore, the probability of flip bits on a specific row depends on the data that are kept in this row as well as the orientation of the specific row cells [5]. If the orientation of the modules is not known before the attack then a different data pattern test must be realized in order to check for the rowhammer vulnerability on specific system [5,6,14].

3. Rowhammer Attack

Hardware fault-attacks typically need to expose the device to specific physical conditions outside of its specification which require physical access to the device under attack. However, through rowhammer attack hardware faults are induced by software without a need to have direct physical access to the system.

In this section, the challenges of triggering rowhammer in a security-relevant manner will be described by summarizing existing exploitation techniques under different architectures, OS (Operating System) and already deployment countermeasures. To perform a successful end-to-end rowhammer attack the following four steps must be implemented/followed.

3.1. Determine the Device under Attack Specific Memory Architecture Characteristics

For probabilistic rowhammer induction approaches, it is enough to know the *row size* so that addresses on the same banck can be found [10]. But the physical address of the bits that correspond to each row, bank, DIMMs (Dual In-line Memory Module) and Memory channel is crucial to perform deterministic *double-side rowhammer*. This DRAM address mapping schemes used by the CPUS's memory controllers is not publicly known by major chip companies like Intel and ARM. Reverse engineering of the *physical address mapping* is proposed for Intel Sandy Bridge CPUs in [15] and used for the *double-side rowhammer* approach in [14]. The DRAMA paper [8] presented two other approaches in order to solve the issue: the first one uses physical probing of the memory bus through a high-bandwidth oscilloscope by measuring the voltage on the pins at the DIMM slots, the second uses *time analysis* that is based on the *rowbuffer conflict* so that it can find out address pair belonging to the same bank and then use this address set to reconstruct the specific map function automatically. Similarly, the authors in [16] are presenting their own specific algorithm based too on timing analysis based approach. The DRAMA solution with exhaustive search avoids the need to know the complex logic behind the memory addressing with a cost that is smaller than the algorithm proposed in [16]. Note that both approaches can be used on Intel or ARM platforms.

3.2. Managing to Activate Rows in Each Bank Fast Enough to Trigger the Rowhammer Vulnerability

Various methods have been used to bypass, clean or nullify all levels of cache in order to have direct access to the physical memory. The eviction of cache lines corresponding to the aggressor row addresses between its memory accesses has been approached in many ways. The first research approaches use directly the CFLUSH command available on user space (userland) for x86 devices [5,10,14]. Early software level countermeasures regarding the above approach were described in [10]. However, such countermeasures were overpassed and new attacks were implemented on different attack interfaces such as scripting language based attacks using web browsers that are triggered remotely [17]. In line with this research direction, new, different, eviction strategies have been proposed as a replacement for the flush instructions. The aim of such strategies is to find an eviction set that contains addresses belonging to the same cache set of the aggressor rows. Such addresses are accessed one after the other so as to force the eviction of our aggressor row. This can be achieved through the use of a timing attack to find out the eviction set [12,17,18], or by using reverse engineering study of the Device under attack, focusing on manipulating the complex hash functions used by modern Intel processor to further partition the cache into slices [9]. On another approach proposed in [1], rowhammer is triggered with non-temporal store instructions, specifically *libc* functions, taking advantages of its cache bypass characteristic. In the most recent studies under ARM architecture, the use of the Direct Memory Access memory management mechanism is proposed as the best way of bypassing CPUs and their caches [7].

3.3. Access the Aggressor Physical Address from Userland

In many cases access mechanisms, memory physical addresses from OS userland is a non-trivial action because generally, the unprivileged processes use virtual memory address space and the virtual to physical map interface is not public knowledge or has been no longer allowed in many OS versions after 2015 (acting as a rowhammer countermeasure) [19]. Attacks that was mounted after consulting their own pagemap interface to obtain the corresponding physical address of each element have been relegated to be applicable just on Linux kernel versions at most 4.0 [12,18]. Exploitation technics based on probabilistic approaches that just pick a random virtual address to trigger the bug [10,11] are based on test code [14] and on the native code of the work in [17]. On the other hand, physical addresses can be accessed through the use of huge virtual pages that are backed by contiguous physical addresses which can use relative offsets to access specific physical memory pages [20]. Note that for the rowhammer bug triggered from a scripting language (e.g., JavaScript) large type arrays that are allocated on 2MB regions are being used and a tool that translates JavaScript arrays indices to physical addresses is required to trigger the bit flips on known physical memory locations [9,17].

3.4. Exploit the Rowhammer Vulnerability (Take Advantage of the Bit Flips)

In the attack of [10], a probabilistic exploitation strategy that sprays the memory with page tables is proposed, hoping that at least one of them lands on a vulnerable region. A second probabilistic exploitation uses a timing-based technique to learn in which bank secret data are kept in order to flip their bits and then just trigger the bug on the specific bank and hope to produce a flip bit of the row's secret data [18]. Other attacks rely on special memory management features such as memory deduplication [20], where the OS can be tricked to map a victim owner memory page on the attacker-controlled physical memory pages, or MMU (Memory Management Unit) Paravirtualization [16], where a malicious VM is allowed to trick a VM (Virtual Machine) monitor into mapping a page table into a vulnerable location. Also, specific target secret can be found and the victim component can be tricked to place it on a vulnerable memory region on deterministic approaches. Finally, a technique is proposed in [7] that is based on predictable memory reuse patterns of standard physical memory allocator.

3.5. Attacks on x86 Devices

Various exploitation techniques have been implemented that exploit the rowhammer vulnerability aiming at bypassing all current defenses and completely subvert the x86 systems.

The x86 instruction set architecture has been implemented in processors from Intel, Cirix, AMD, VIA and many other companies and is used in several new generations of embedded systems. Rowhammer attacks against such devices aiming directly in memory accesses and not the OS itself are hard to detect with traditional software tools like antimalware or Firewall products. The potential diversity of the attack can lead to a broad range of possible compromises including unauthorized accesses, denial of services, or cryptography key theft. Rowhammer attacks can also be used as a backdoor (an attack vector) for more complex attacks that would otherwise be detected and failed in x86 architectures.

Several Rowhammer attacks exploiting x86 architecture causing serious security breaches were described in [12]. In one such attack, by triggering Rowhammer vulnerability bit flips are caused in Native Client (NaCl) program thus achieving privileges escalation, escaping from the x86 NaCl and acquiring the ability to make OS system calls (syscalls).

In a different attack [10], physical memory is filled with page tables for a single process (nmap() function is used to repeatedly fill a file's data in memory). This mapping sprays the memory with page table entries (PTEs), that are used to translate the newly mmap()'ed virtual addresses [12]. Performing rowhammer in memory containing page tables, there is a non-trivial probability that some PTE through bit flipping will be changed so that it points to a physical page that contains a page table. This will give the application we are using access to its own page tables thus giving to the application user (i.e., the attacker) full Read/Write permission to a page table entry, an action that results eventually to access the full physical memory [10,12].

Rowhammer presents a purely software driven way of inducing fault injection attacks [21] for finding private keys by flipping bits in memory. Such an attack is mounted for RSA (Rabin Shamir Adleman) exponentiation where a bit flip is injected through rowhammer in the secret exponent [18] This presents a new threat to the security of the embedded systems because it unlocks a hardware attack type (fault injection attacks) that traditionally need physical access to devices. Based on the theory that a single fault signature is enough to leak the secret in an unprotected implementation, the attacker aims at bit flips on the exponent, however having user-level privileges on the system, he does not know where the secret is placed in LLC (Last Level Cache) and DRAM. Thus, the attacker needs to identify the bank in which the secret exponent resides. This goal in [18], is achieved through a spy process which uses timing analysis to identify the channel, bank and rank where the secret is mapped by probing on the execution time of the RSA decryption algorithm. Then, the *eviction set* that corresponds to the secret set must be found so as to bypass cache without flush instructions. Finally, the Rowhammer vulnerability can be triggered on the bunk, where the public key is stored, thus resulting, with high probability of a bit flip of this key. Then performing a fault injection analysis will reveal from the fault public key, the secret exponent key.

The exploit of cross-VM rowhammer attacks was proven to be successful in server-side machines by achieved private key exfiltration from HTTPS web server and code injection that can bypass password authentication on an OpenSSH (SSH: Secure Shell) server [16]. This lead to a serious threat on multi-tenant infrastructure clouds since user clouds may collocate their virtual machines on the same physical server sharing hardware resources, including DRAMS. The attacks use paravirtualization VM to break VM isolation and compromise the integrity and confidentially of co-located VMs. The vulnerability relies on the fact that the mapping between the pseudo-physical memory page to machine memory for the user VM space can be tricked to point to memory pages of other VM or even the hypervisor. The presented technique, called *page table replacement attack*, conducts rowhammer attacks to flip bits deterministically in a page directory entry (PDE), thus making in a pointer to a different page table.

Abusing the Linux's memory deduplication system (KSM) and using the rowhammer vulnerability, an attacker can reliably flip bits in any physical page of the software stack [20]. A VM can abuse memory deduplication to place on a control target physical page a co-hosted victim VM page and then exploit the rowhammer to flip a particular bit in the victim page without permission to write on that bit. Due to the high popularity of KSM in production clouds, this technique can have devastating consequents in common cloud settings.

A remote software-induced rowhammer fault injection has been implemented on script languages [17]. The attack is executed in sandbox JavaScript, that is enabled by default on every modern browser, therefore can be launched from any website. The specific implementation is independent of the instruction set of the CPU and its main challenge relies on performing an optimal cache eviction strategy as a replacement for the flush instruction. Another two different thread models are proposen [11] based on x86 non-temporal store method. They are very convenient because they can be employed to implement attacks with the widely used memset and memcpy functions from libc library, thus giving to almost every code the potential of rowhammering. One of the models can be used to implement a code from untrusted source and scape NaCl sandbox. The second model can be used to trigger rowhammer with existing benign code that does not compromise the security, for example, multimedia players or PDF readers by compromising inputs.

Both, script and native, language-based approaches mentioned above, achieve to trigger rowhammer on remote systems and constitute a basis for adapting the existing vulnerability exploitation techniques to gain root privileges with a remote attack and access all the physical memory of a target system. Both can beat the basic rowhammer countermeasure since they bypass the flush instruction (banned in some latest OS versions). The second approach follows a very simple attack methodology, while the JavaScript approach is the easier to be executed by a remote attacker.

3.6. Attack on ARM Architecture

Deterministic rowhammer attacks are feasible on commodity mobile platforms by using the end-to-end DRAMMER attack on Android [7]. In this work, the Android/ARM exploitation that was employed in order to mount Rowhammer attack was Google's ION memory management tool that provides DMA buffer management API's (Application Programming Interface) from user space. Modern computing platforms need to support efficient memory sharing between their several different hardware components as well as between devices and userland services. Thus, the OS must provide allocators that support operations to physical contiguous memory pages since most devices need this kind of DMA (Direct Memory Access) performance. The DMA buffer management API provides from userland uncached memory access to physical memory addressing directly to the DRAM, solving two of the more complex Rowhammer attack requirements (steps 1 and 3).

More specifically, in the DRAMMER attack approach, initially predictable reuse patterns of standard physical memory allocator are used in order to place the target security-sensitive data at a vulnerable position in the physical memory [20]. Then, a memory template is created by triggering rowhammer on a big part of the physical memory, looking for bit flips. Afterward, the vulnerable location where the secret is placed is chosen and then *memory massaging* is performed by exhausting available memory chunks of varying sizes so as to drive the physical memory allocator into a state that brings the device under attack in the specific vulnerable memory region that we have predicted.

The goal of the DRAMMER ARM-based attack in [7] is to root the android device. This is achieved by overwriting the control page table to probe kernel memory from user *struct cred* structures [7]. The empirical studies provided in [7] points out that this end-to-end attack takes from 30 s to 15 min on the most vulnerable phone. However, the most important conclusion from the attack is that it exposes the vulnerability of many of the tested ARMv7 devices to the rowhammer bug which was considered not possible till now thus opens the road for rowhammering a wide variety of ARM-based embedded systems that have DRAM.

This final conclusion has several repercussions on the security of the latest IT enabled embedded system devices that still have small memory capabilities. This also includes mobile phones that in some cases can be used for fast incident response (e.g., in crisis management or critical infrastructure incidents). As described in the previous paragraphs, mobile devices can be vulnerable to rowhammer based attacks so they constitute an attack entry point to the security critical system.

Apart from this, embedded system device can also be rowhammer attacked. Most of them do not rely on ARM processors but this is gradually changing as ARM is dominating the marker and provides security benefits like ARM Trustzone. Also, most of them, currently, rely on SRAM which is hard to rowhammer, but there also exist embedded system devices with high processing power like the Raspberry Pi that relies on LPDDR3 (Low Power DRAM has also been found to be susceptible to the rowhammer bug) and constitutes an interesting rowhammer attack target. Furthermore, there is the embedded system approach from Intel (Intel Edisson, Intel Galileo) that has very attractive performance capabilities for embedded system world applications and Internet of Things (IoT). Note that Intel Edisson or Galileo rely on the x86 instruction set Architecture and thus they are easily rowhammmer attacked especially since they also have LPDDR3.

As the above embedded solutions are bound to be integrated also in the new real-time system end nodes, like those of a critical infrastructure (e.g., a power plant, an airport, a bridge) the rowhammer attack vulnerability should be taken into account in choosing the appropriate device for deployment in the field.

4. Side-Channel Analysis Attacks in Embedded System Devices

Side-channel analysis attacks exploit a device under attack hardware characteristics leakage (power dissipation, computation time, electromagnetic emission etc.) to extract information about the processed data and use them to deduce sensitive information (cryptographic keys, messages etc.). An attacker does not tamper with the device under attack in any way and needs only make appropriate observations to mount a successful attack. Such observation can be done remotely or physically through appropriate tools. Depending on the observed leakage, the most widely used SCAs are microarchitectural/cache, timing, power dissipation, electromagnetic Emission attacks.

4.1. Microarchitectural/Cache Attacks

Microarchitectural side channel attacks exploit the operation performed in a computer architecture during software execution. Fundamentally, a computer architecture design aims at optimizing processing speed, thus computer architects have designed machines where processed data are strongly correlated with memory access and execution times. This data and code correlation acts as an exploitable side channel for Microarchitectural SCAs. Since cache is strongly related to execution/processing time, has high granularity and is lacking any access restrictions, it constitutes an ideal focal point for Microarchitectural SCAs. Microarchitectural attacks are impervious to access boundaries established at the software level, so they can bypass many cyberattack application layer software countermeasure tools and have managed to be effective even against restrictive execution environments like Virtual Machines (VMs) [22] or ARM TrustZone [23].

As in all SCAs, cache-based attacks aim at recovering secret keys when executing a known cryptographic algorithm. Typically, there exist access based cache attacks where an attacker extracts results from observing his own cache channel (measuring cache access time) and timing based cache attacks where the attacker extracts results from measuring a victim's cache hits and misses to infer access time [24]. In [24,25], the authors provide a more detailed categorization of four attack types. Type I attacks on Cache Misses due to external interference, Type II attacks on Cache Hits due to external interference kind of attacks, Type III attacks on Cache Misses due to internal interference, and Type IV attacks on Cache Hits due to internal interference. Interference can be considered external when it is present between the attacker's program and the victim's program while internal is the interference that is related only with the victim's program. Most Type I and II attacks are access based attacks while most Type II and IV attacks are timing based attacks.

One of the first cache attacks was proposed and implemented by Bernstein [26] on AES (Advanced Encryption Standard) algorithm where microarchitectural timing differences when accessing various positions of a loaded in memory AES SBox look up table, were exploited. At the same year, another access based attack from Percival was also proposed in [27] focusing on openSSL key theft and RSA. Later, the PRIME + PROBE and EVICT + TIME attack approaches were introduced in [28]. Both attacks relied on the knowledge of a cache state before an encryption and the capturing of the cache changes during encryption to deduce sensitive data. Cache attacks were also expanded so as to compromise public key cryptography algorithms like RSA as proposed in [27] and later in [29], demonstrating that such attacks are possible on the full spectrum of popular cryptography algorithms. Attacks became more potent after the proposal of the FLUSH + RELOAD attack, described in [30,31] which exploits the shared memory pages of OS libraries stored in the Last Level Cache (LLC) of any computer and similarly to sophisticated variations of the PRIME+PROBE attack [32] also focused on LLC, became applicable in cross core applications even against VM devices [22,32,33]. Furthermore, variations of the FLASH + RELOAD attack have been proposed for ARM-based systems thus providing strong implications of cache SCA vulnerabilities in ARM embedded systems (including embedded system nodes or Android-based mobile devices systems and ARM TrustZone Enabled processes) [23,34,35].

Through the relevant research history, microarchitecture SCAs have exploited several different covert channels existing in typical computer systems. L1 Cache is one of the main targets of microarchitecture/cache SCAs either holding data (data L1 cache) or instructions (instruction L1 cache). The small size of this cache type, as well as the separation between data and instructions, can help an attacker to monitor the entire cache and relate the collected leakage to specific data or instruction blocks. On the other hand, the fact that L1 (and L2) cache in modern multicore processor systems is core-private, makes SCAs related to L1 very difficult in such systems.

Another convert channel for microarchitecture/cache SCAs is related to the Branch Processing Unit (BPU). Observing if there are branch mispredictions (through time differences between correct and wrong predictions) can be also a way of inducing an attack. However, the BPU is not easily accessible to modern processors and the above mentioned time differences are so small that is hardly observable. Also, BPU is core-private thus fairs the same problems as L1 cache channel [22].

The LLC constitute the most advantageous channel for mounting microarchitecture/cache SCAs because it is shared between processor cores and is associated with the processor main memory in a way that enables the discrimination of LLC accesses from memory accesses with small error rate (due to significant time differences) [22,30].

4.2. Other SCAs: Power Analysis/Electromagnetic Emission Attacks

Apart from the above microarchitectural/cache SCAs for the shake of completeness it should also be mentioned that there exist a very broad variety of other SCAs that require physical contact or close proximity to a device under attack in order to be mounted successfully. This scenario is not unrealistic in the case of embedded systems since there exist many in field deployed such devices that can be physically accessed by an attacker without detection (in industrial areas, in critical infrastructures systems). Such SCAs are very potent and can be successful even against dedicated hardware security modules if the latter do not have some relevant installed countermeasure. Typically, the goal of the attacks is to retrieve cryptographic keys that are processed in a target device under attack from the leakage trace that is emanated during this execution. Apart from timing based SCAs that were already discussed in the previous subsections, the most popular SCAs are the ones exploiting the power dissipation of a target device, or the Electromagnetic emission (EM) that is leaking from the target device. Both attacks are of similar nature since their leakage is related to current fluctuations happening during a cryptographic operation execution.

To model SCAs we can adopt the approach described in [36,37]. Assume that we have a computation C that consists of series of O_0 or O_1 operations that require inputs X_0 and X_1 respectively (thus O_i (X_i) for i in {0, 1}). During processing of the C computation, each operation can be linked to an information leakage variable L_i. A side channel analysis attack is possible if there is some secret information s that is shared between O_i and its leakage L_i. The ultimate goal of a side channel analysis is, by using an analysis strategy, to deduce s (secret value) from the information leakage L_i. The simplest way to achieve that is by examining a sequence of O_i operations in time to discover s. Simple SCAs (SSCAs) can be easily mounted in a single leakage trace (e.g., in RSA of ECC (Error Correction Codes) implementation) and are typically horizontal type of attacks meaning that they are mounted using a single leakage trace processed in time. When SSCAs are not possible, advanced SCAs (ASCAs) must be mounted to extract s [36].

Advanced SCAs do not focus only on the operations (e.g., O_i) but also on the Computation operands [36,37]. Advanced SCAs are focused on a subset of the calculation C (and/or O_i) and through collection of sufficiently large number N of leakage traces $L_i(t)$ for all t in {1, ..., N} using inputs $X_i(t)$ exploit the statistical dependency between the calculation on C for all X_i and the secret s. ASCAs follow the hypothesis test principle [36,37] where a series of hypothesis s' on s (usually on some byte or bit j of s i.e., $s_j' = 0$ or 1) is made and a series of leakage prediction values are found based on each of these hypothesis using an appropriate prediction model. The values of each hypothesis are evaluated against all actual leakage traces using an appropriate distinguisher Δ for all inputs X_i so as to decide which hypothesis is correct.

SSCAs and ASCAs can follow either the vertical or horizontal leakage collection and analysis strategies. In the vertical approach, the implementation is used N times employing either the same or different inputs each time t in order to collect traces-observations $L_i(t)$. Each observation is associated with t-th execution of the implementation. In the horizontal approach, leakage traces-observations are collected from a single execution of the implementation under attack and each trace corresponds to a different time period within the time frame of this execution. As expected, in Horizontal attacks the implementation input is always the same. ASCAs are also called and Differential SCAs.

The distinguisher used in order to evaluate the hypothesis against the actual leakage trace is usually a statistical function like Pearson correlation or collision correlation.

5. Countermeasures

5.1. Rowhammer Attacks Countermeasures

There exist various countermeasures for rowhammer attacks proposed in the literature. In [5] various system level mitigations such as enforcing Error Correction Codes (ECC) protection, increasing the refresh rate of DRAM cells, retiring DRAM cells that the DRAM manufacturer identifies as victim cells and refreshing vulnerable rows were proposed.

With the exposure-publication of the rowhammer vulnerability attack, specific software based solutions were introduced for mitigating the problem. The first such countermeasure was the disallowance or rewriting of the flush instructions, that have been deployed in Google NaCl. Also, on the new kernel Linux, the pagemap interface is prohibited from userland so as to block relevant rowhammer attacks [19]. However, new attacks have appeared that overpassed the above countermeasure, by providing rowhammer triggering using JavaScrip [17] or native code [11], without needing cache eviction. Similarly, the pagemap based countermeasure has been bypassed by the use of different methodologies of finding out the machine address map to a physical address or the map to the virtual userland [8,14,16].

Furthermore, some hardware manufacturers have implemented a series of measures to reduce the probability of a successful attack. However, the solution of providing memories with ECC does not offer strong security guarantees [6] because ECC cannot detect multiple bit flips on the same row in order to correct them. Most of the hardware based countermeasure just double the DRAM refresh rate however even then the probability of bit flips is still high enough to be exploited in some machines because the rate will need to be increased up to eight times to achieve a competent mitigation [5].

There also exist more generic countermeasures that try to detect rowhammer by monitoring the last-level cache misses on a refresh interval and row access with high temporal locality on Intel/AMD processors [12]. If the missing cache overpasses a threshold, a selective refresh is performed on the vulnerable rows.

Other countermeasures rely on detection of activation patterns. Following this direction, in [5], the PARA (Probabilistic Adjacency Row Activation) mechanism is implemented in the memory controller. This countermeasure's approach is to refresh the adjacent row of the accessed row with a probability p each time (acts not deterministically) thus avoiding the hardware costly complex data structure for counting the number of row activations to mitigate the rowhammer attack. Similarly, another memory controller implementation solution is ARMOR [38] that includes a cache buffer for frequently activated rows.

DDR4 memory modules are harder to rowhammer since they include Targeted Row Refresh (TTR) techniques in the form of a special module tracking the row activation frequency so as to select vulnerable rows for refreshing [39]. The LPDDR4 standard also incorporates TTR as well as the Maximum Active Count (MAC) technique [40]. Yet still, as noted in [8] the rowhammer vulnerability bug exists in DDR4.

When there is a need for a high-security level in embedded system devices (e.g., in critical infrastructures) the above-described countermeasures are not enough to guarantee this high level. More specialized solutions must be developing for each kind of architecture, OS and usage in order to shield the devices from the exploited vulnerabilities used in the existing rowhammer attacks. The first step in this process is to develop a mechanism to disable memory deduplication.

5.2. Side Channel Analysis Attack Countermeasures

As was described above, SCAs can be very potent if appropriate countermeasures are not introduced in the device under attack. All SCAs rely on some side channel information leakage, so SCA countermeasures aim at either hiding this leakage or at minimizing this leakage so that it contains very small or no useful information to mount an attack (making the leakage trace, data independent) [41]. The most widely used countermeasure, primarily applicable on microarchitecture attacks, is making the security operation time delay constant or random, regardless of the microarchitecture elements that are used or the implemented code it is executed [42]. However, developing constant-time execution code is not always easy since optimizations introduced by the compiler must be bypassed. For this reason, specialized constant-time libraries have been made available in the literature, to help security developers protect their applications against SCAs. Similar to constant time implementations, effort has also been invested to structure power and EM constant approaches. Such approaches follow algorithmic modifications of the cryptography functions realized either in software or in hardware [36]. In the latter case, circuit based modifications can also be made so as to thwart SCAs (e.g., Dual Rail Technology, special CMOS circuits) [43]. The goal of these countermeasures is making the leakage trace of O_1 indistinguishable from the leakage trace of O_0. Hardware support has also been introduced in certain processor technologies (e.g., Intel processors) to make constant-time cryptography operations (e.g., for AES).

For microarchitectural/Cache SCAs, cache specific countermeasures have been introduced in literature. A good overview of the countermeasure directives is made in [38] and [24]. Typically, countermeasures can be divided in isolation measures and randomization measures. By isolation, the attacker no longer shares cache with the victim so external interference related attacks become very difficult. Isolation can be achieved by splitting cache into zones related to individual processed for example using hardware virtualization. This partitioning can be done either statically or dynamically. Randomization is a more generic countermeasure that is applicable in most of the SCAs. In practice, it can be supported by random cache eviction or random cache permutation mechanisms [24].

In general, the provided solutions, try to nullify the cache characteristics that cache SCAs exploit (e.g., FLUSH + RELOAD, PRIME + PROBE and EVICT + TIME). Among the existing solutions (fitted in the above-mentioned categories) there exist techniques to eliminate timing side channels by introducing virtual time and black-box mitigation techniques [42]. Also, time partitioning can be used through regular Cache Flushing, Lattice Scheduling, Memory Controller Partitioning, structuring Execution Leases and performing Kernel address space isolation. Finally, hardware partitioning is a proposed countermeasure approach that is focused on disabling hardware threading, page sharing, introducing Hardware Cache Partitions, quasi-partitioning even migrating VMs of cloud services. There also exist approaches that by modeling a cache memory's side channel leakage, they can evaluate a system's vulnerability to specific cache side channel attack categories [24] thus constituting a proactive cache SCA countermeasure.

It must be mentioned here, that some of those countermeasures can be bypassed using the rowhammer attacks using the approaches described in previous subsections.

Apart from cache SCAs countermeasure, randomization is a favorable solution for countering all ASCAs (both horizontal and vertical). Using randomization, the sensitive information is disassociated from the leakage trace and is hidden by multiplicatively or additively blinding this information using a randomly generated number. A variant of this technique is information hiding where random noise is added to the leakage channel thus scrabbling making the associate to the leakage trace unusable [36].

6. Conclusions

In this paper, a survey was made on malicious attacks that can be mounted on real-time embedded system devices due to computer architecture or hardware vulnerabilities and side channel leakage exploitation. Such attacks use software code that does not directly interfere with other software tools (infect programs, directly alter their functionality) but rather exploit intrinsic vulnerabilities on the embedded system device hardware to give leverage to an attacker. In this paper, we focus on the recently exploited rowhammer vulnerability and on SCAs attacks (microarchitectural/cache attacks and power/EM analysis attacks) and provide an overview of the attack research directions highlighting their applicability in the various computer architectures. The attacks' analysis performed in this paper indicate that the mentioned attacks can reduce the QoS level of an embedded system as well as defeat many security and trust enhancing technologies including software cyberattack tools (antimalware, anomaly detection, firewalls) but also hardware assisted technologies like memory isolation through VMs, Trusted Execution Environments, ARM TrustZone or dedicated security elements. The real-time embedded system security architect must be informed of this attack taxonomy and must put pressure on embedded system device manufacturers to provide security patches to their products, capable of thwarting the attacks.

Acknowledgments: The work in this paper is supported by CIPSEC EU Horizon 2020 project under grant agreement No. 700378.

Author Contributions: Apostolos P. Fournaris conceived and structured the paper's concept, did the overview on cache35 attacks, side channel attacks and countermeasures and contributed to the survey research of the rowhammer microarchitectural attacks, overviewed the whole text and wrote Sections 4 and 5.2. Lidia Pocero Fraile did the overview research for the Rowhammer attack and was responsible for Sections 2 and 3. Odysseas Koufopavlou overviewed the research process, contributed to the paper's concept and overall text formation.

Conflicts of Interest: The authors declare no conflict of interest. The founding sponsors had no role in the design of the study; in the collection, analyses, or interpretation of data; in the writing of the manuscript, and in the decision to publish the results.

References

1. Enhancing Critical Infrastructure Protection with Innovative SECurity Framework (CIPSEC). H2020 European Project. Available online: www.cipsec.eu (accessed on 28 March 2017).
2. Challener, D.; Yoder, K.; Catherman, R.; Safford, D.; Van Doorn, L. *A Practical Guide to Trusted Computing*; IBM Press: Indianapolis, IN, USA, 2007.
3. Trusted Computing Group. *TCG TPM Specification Version 2.0*; Trusted Computing Group: Beaverton, OR, USA, 2014.
4. ARM. ARMTrustZone. Available online: https://www.arm.com/products/security-on-arm/trustzone (accessed on 1 April 2017).
5. Kim, Y.R.; Daly, J.; Kim, C.; Fallin, J.; Lee, H.; Lee, D.; Wilkerson, C.; Lai, K.; Mutlu, O. Flipping bits in memory without accessing them: An experimental study of DRAM disturbance errors. In Proceedings of the ACM/IEEE 41st International Symposium on Computer Architecture (ISCA), Minneapolis, MN, USA, 14–18 June 2014.
6. Lanteigne, M. How Rowhammer Could Be Used to Exploit Weakness in Computer Hardware. 2016. Available online: https://www.thirdio.com/rowhammer.pdf (accessed on 1 April 2017).
7. Van der Veen, V.; Fratantonio, Y.; Lindorfer, M.; Gruss, D.; Maurice, C.; Vigna, G.; Bos, H.; Razavi, K.; Giuffrida, C. Drammer: Deterministic rowhammer attacks on mobile platforms. In Proceedings of the 2016 ACM SIGSAC Conference on Computer and Communications Security (CCS'16), Vienna, Austria, 24–28 October 2016.
8. Pessl, P.; Gruss, D.; Maurice, C.; Schwarz, M.; Mangard, S. DRAMA: Exploiting DRAM addressing for Cross-CPU attacks. In Proceedings of the USENIX Security Symposium, Austin, TX, USA, 10–12 August 2016.
9. Bosman, E.; Razavi, K.; Bos, H.; Giuffrida, C. Dedup Est Machina: Memory Deduplication as an Advanced Exploitation Vector. In Proceedings of the 2016 IEEE Symposium on Security Privacy, SP 2016, San Jose, MA, USA, 23–25 May 2016; pp. 987–1004.

10. Seaborn, M.; Dullien, T. Exploiting the DRAM rowhammer bug to gain kernel privileges. In Proceedings of the 2016 ACM SIGSAC Conference, Vienna, Austria, 24–28 October 2016.

11. Qiao, R.; Seaborn, M. A new approach for rowhammer attacks. In Proceedings of the 2016 IEEE International Symposium on Hardware Oriented Security and Trust (HOST), McLean, VA, USA, 3–5 May 2016.

12. Aweke, Z.B.; Yitbarek, S.F.; Qiao, R.; Das, R.; Hicks, M.; Oren, Y.; Austin, T. ANVIL: Software-based protection against next-generation rowhammer attacks. In Proceedings of the 21st ACM International Conference on Architectural Support for Programming Languages and Operating Systems (ASPLOS), Atlanta, GA, USA, 2–6 April 2016.

13. Moinuddin, K.Q.; Dae-Hyun, K.; Samira, K.; Prashant, J.N.; Onur, M. AVATAR: A variable-retention-time (vrt) aware refresh for dram systems. In Proceedings of the IEEE/IFIP International Conference on Dependable Systems and Networks (DSN), Rio de Janeiro, Brazil, 22–25 June 2015.

14. Program for Testing for the DRAM Rowhammer Problem. 2015. Available online: https://github.com/google/rowhammer-test (accessed on 15 March 2017).

15. Seaborns, M. How Physical Adressesses Map to Rows and Banks in DRAM. 2015. Available online: http://lackingrhoticity.blogspot.com/2015/05/how-physical-adresses-map-to-rows-and-banks.html (accessed on 5 April 2017).

16. Xiao, Y.; Zhang, X.; Teodorescu, R. One bit flips, one cloud flops: Cross-VM row hammer attacks and privilege escalation. In Proceedings of the 25th USENIX Security Symposium, Austin, TX, USA, 10–12 August 2016.

17. Gruss, D.; Maurice, C.; Mangard, S. Rowhammer.js: A remote software-induced fault attack in javascript. In Proceedings of the 13th Conference on Detection of Intrusions and Malware Vulnerability Assessment (DIMVA), Donostia-San Sebastián, Spain, 7–8 July 2016.

18. Bhattacharya, S.; Mukhopadhyay, D. Curious Case of Rowhammer: Flipping Secret Exponent Bits Using Timing Analysis. *Lect. Notes Comput. Sci.* **2016**, *9813*, 602–624.

19. Salyzyn, M. AOSP Commit 0549ddb9: UPSTREAM: Pagemap: Do Not Leak Physiscal Addresses to Non-Privilahe Userspace. 2015. Available online: http://goo.gl/Qye2MN (accessed on 1 May 2017).

20. Razai, K.; Gras, B.; Bosman, E.; Preneel, B.; Giuffrida, C.; Bos, H. Flip feng shui: Hammering a needle in the software stack. In Proceedings of the 25th USENIX Security Sympoisium, Austin, TX, USA, 10–12 August 2016.

21. Joye, M.; Tunstall, M. *Fault Analysis in Cryptography*; Springer: New York, NY, USA, 2012.

22. Irazoqui, G.; Eisenbarth, T.; Sunar, B. Cross processor cache attacks. In Proceedings of the 2016 ACM Asia Conference Computer Communications Security, Xi'an, China, 30 May–3 June 2016; pp. 353–364.

23. Zhang, N.; Sun, K.; Shands, D.; Lou, W.; Hou, Y.T. *TruSpy: Cache Side-Channel Information Leakage from the Secure World on ARM Devices*; Cryptology ePrint Archive Report 2016/980; The International Association for Cryptologic Research (IACR): Cambridge, UK, 2016.

24. Zhang, T.; Lee, R.B. *Secure Cache Modeling for Measuring Side-Channel Leakage*; Technical Report; Princeton University: Princeton, NJ, USA, 2014; Available online: http://palms.ee.princeton.edu/node/428 (accessed on 3 March 2017).

25. Wang, Z. Information Leakage Due to Cache and Processor Architectures. Ph.D. Thesis, Princeton University, Princeton, NJ, USA, 2012.

26. Bernstein, D.J. Cache-Timing Attacks on AES. 2005. Available online: https://cr.yp.to/antiforgery/cachetiming-20050414.pdf (accessed on 12 April 2017).

27. Percival, C. Cache missing for fun and profit. In Proceedings of the BSDCan 2005, Ottawa, ON, Canada, 13–14 May 2005.

28. Osvik, D.A.; Shamir, A.; Tromer, E. Cache attacks and countermeasures: the case of AES. In *Topics in Cryptology–CT-RSA 2006*; Springer: New York, NY, USA, 2006; pp. 1–20.

29. Aciicmez, O.; Brumley, B.B.; Grabher, P. New results on instruction cache attacks. In Proceedings of the 12th International Conference on Cryptographic Hardware and Embedded Systems, Santa Barbara, CA, USA, 17–20 August 2010; pp. 110–124.

30. Yarom, Y.; Falkner, K. Flush+ reload: A high resolution, low noise, l3 cache side-channel attack. In Proceedings of the 23rd USENIX Security Symposium (USENIX Security 14), Santa Barbara, CA, USA, 20–22 August 2014; pp. 719–732.

31. Gullasch, D.; Bangerter, E.; Krenn, S. Cache games–bringing access-based cache attacks on AES to practice. In Proceedings of the 2011 IEEE Symposium on Security and Privacy (SP), Berkeley, CA, USA, 22–25 May 2011; pp. 490–505.
32. Liu, F.; Yarom, Y.; Ge, Q.; Heiser, G.; Lee, R.B. Last-level cache side-channel attacks are practical. In Proceedings of the 2015 IEEE Symposium on Security and Privacy, San Jose, CA, USA, 17–21 May 2015; pp. 605–622.
33. Irazoqui, G.; Inci, M.S.; Eisenbarth, T.; Sunar, B. *Wait a Minute! A Fast, Cross-VM Attack on AES*; Lecture Notes Computer Science; Springer: New York, NY, USA, 2014; pp. 299–319.
34. Zhang, X. Return-oriented flush-reload side channels on arm and their implications for android security. In Proceedings of the 2016 ACM SIGSAC Conference on Computer Communications Security (CCS'16), Vienna, Austria, 24–28 October 2016; pp. 858–870.
35. Lipp, M.; Gruss, D.; Spreitzer, R.; Maurice, C.; Mangard, S. ARMageddon: Cache attacks on mobile devices. In Proceedings of the 25th USENIX Security Symposium, Austin, TX, USA, 10–12 August 2016; pp. 549–564.
36. Fournaris, A.P. Fault and power analysis attack protection techniques for standardized public key cryptosystems. In *Hardware Security and Trust: Design and Deployment of Integrated Circuits in a Threatened Environment*; Sklavos, N., Chaves, R., di Natale, G., Regazzoni, F., Eds.; Springer: Cham, Switzerland, 2017; pp. 93–105.
37. Bauer, A.; Jaulmes, E.; Prouff, E.; Wild, J. Horizontal and vertical side-channel attacks against secure rsa implementations. In *Topics in Cryptology—CT-RSA 2013: The Cryptographers' Track at the RSA Conference 2013*; Dawson, E., Ed.; Springer: Berlin/Heidelberg, Germany, 2013; pp. 1–17.
38. Ghasempour, M.; Lujan, M.; Garside, J. ARMOR: A Run-Time Memory Hot-Row Detector. 2015. Available online: http://apt.cs.manchester.ac.uk/projects/ARMOR/Rowhammer (accessed on 6 March 2017).
39. *DDR4 SDRAM MT40A2G4, MT401G8, MT40A512M16 Datasheet, 2015*; Micro Inc.: Irvine, CA, USA, 2015.
40. JEDEC Solid State Techonlogy Association. *Low Power Double Data Rate 4 (LPDDR4)*; JEDEC Solid State Techonlogy Association: Arlington, VA, USA, 2015.
41. Mangard, S.; Oswald, E.; Popp, T. *Power Analysis Attacks: Revealing the Secrets of Smart Cards (Advances in Information Security)*; Springer: New York, NY, USA, 2007.
42. Ge, Q.; Yarom, Y.; Cock, D.; Heiser, G. A survey of microarchitectural timing attacks and countermeasures on contemporary hardware. *J. Cryptogr. Eng.* **2016**, 1–27. [CrossRef]
43. Tiri, K.; Verbauwhede, I. A digital design flow for secure integrated circuits. In Proceedings of the IEEE Transactions CAD Integrated Circuits System, Pennsylvania, PA, USA, 12–15 August 2006; pp. 1197–1208.

 *electronics*

Article

A Formally Reliable Cognitive Middleware for the Security of Industrial Control Systems

Muhammad Taimoor Khan [1,*], Dimitrios Serpanos [2] and Howard Shrobe [3]

[1] Institute of Informatics, Alpen-Adria University, A-9020 Klagenfurt, Austria
[2] Industrial Systems Institute/RC-Athena & ECE, University of Patras, GR 26504 Patras, Greece; serpanos@ece.upatras.gr
[3] MIT CSAIL, Cambridge, MA 02139, USA; hes@csail.mit.edu
* Correspondence: muhammad.khan@aau.at; Tel.: +43-463-2700-3529

Received: 31 May 2017; Accepted: 8 August 2017; Published: 11 August 2017

Abstract: In this paper, we present our results on the formal reliability analysis of the behavioral correctness of our cognitive middleware ARMET. The formally assured behavioral correctness of a software system is a fundamental prerequisite for the system's security. Therefore, the goal of this study is to, first, formalize the behavioral semantics of the middleware and, second, to prove its behavioral correctness. In this study, we focus only on the core and critical component of the middleware: the execution monitor. The execution monitor identifies inconsistencies between runtime observations of an industrial control system (ICS) application and predictions of the specification of the application. As a starting point, we have defined the formal (denotational) semantics of the observations (produced by the application at run-time), and predictions (produced by the executable specification of the application). Then, based on the formal semantices, we have formalized the behavior of the execution monitor. Finally, based on the semantics, we have proved soundness (absence of false alarms) and completeness (detection of arbitrary attacks) to assure the behavioral correctness of the monitor.

Keywords: run-time monitoring; security monitor; absence of false alarms; ICS; CPS

1. Introduction

Defending industrial control systems (ICS) against cyber-attack requires us to be able to rapidly and accurately detect that an attack has occurred in order to, on one hand, assure the continuous operation of ICS and, on the other, to meet ICS real-time requirements. Today's detection systems are woefully inadequate, suffering from both high false positive and false negative rates. There are two key reasons for this. First, the systems do not understand the complete behavior of the system they are protecting. The second is that they do not understand what an attacker is trying to achieve. Most systems that exhibit this behavior, in fact, are retrospective, that is they understand some surface signatures of previous attacks and attempt to recognize the same signature in current traffic. Furthermore, they are passive in character, they sit back and wait for something similar to what has already happened to reoccur. Attackers, of course, respond by varying their attacks, so as to avoid detection.

ARMET [1] is a representative of a new class of protection systems that employ a different, active form of perception, one that is informed both by knowledge of what the protected application is trying to do and by knowledge of how attackers think. It employs both bottom-up reasoning (going from sensors data to conclusions about what attacks might be in progress) and top-down reasoning (given a set of hypotheses about what attacks might be in progress, it focuses its attention to those events most likely to significantly help in discerning the ground truth).

Based on AWDRAT [2], ARMET is a general purpose middleware system that provides survivability to any kind of new and legacy software system and to ICS in particular. As shown in Figure 1, the run-time monitor (RSM) of ARMET checks consistency between the run-time behavior of the application implementation (AppImpl) and the specified behavior (AppSpec) of the system. If there is an attack, then the diagnostic engine identifies an attack (an illegal behavioral pattern) and the corresponding set of resources that were compromised during the attack. After identifying an attack, a larger system (e.g., AWDRAT) attempts to repair and then regenerate the compromised system into a safer state, to only allow fail-safe, if possible. The task of regeneration is based on the dependency-directed reasoning [3] engine of the system that contributes to self-organization and self-awareness. It does so by recording execution steps intrinsically, the states of the system and their corresponding justification (reason). The details on diagnosis and recovery are beyond the scope of this paper. Based on the execution monitor and the reasoning engine of ARMET, not only is the detection of known attacks possible but also the detection of unknown attacks and potential bugs in the application implementation is possible.

The RSM has been developed as prototype implementations in a general-purpose, computing environment (laptop) using a general-purpose functional programming environment (Lisp). In addition to the software's ability to be easily ported to a wide range of systems, the software can be directly developed in any embedded OS and RTOS middleware environment, such as RTLinux, Windows CE, LynxOS, VxWorks, etc. The current trend toward sophisticated PLC and SCADA systems with advanced OS and middleware capabilities provides an appropriate environment for developing highly advanced software systems for control and management [4].

The rest of this paper is organized as follows: Section 2 presents the syntax and semantics of the specification language of ARMET, while Section 3 explains the syntax and semantics of the monitor. The proof of behavioral correctness (i.e., soundness and completeness) of the monitor is discussed in Section 4. Finally, we conclude in Section 5.

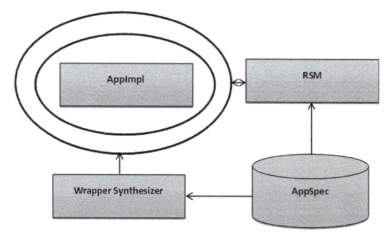

Figure 1. ARMET—Run-time security monitor.

2. A Specification Language of ARMET

A specification language of ARMET allows the description of the behavior of ICS application implementation (AppImpl) based on a fairly high-level description written in a language of "Plan Calculus" [3], which is a decomposition of pre- and post-conditions and invariant for each computing component (module) of the system. The description can be considered as an executable specification of the system. The specification is a hierarchical nesting of a system's components such that the input and output ports of each component are connected by data and control flow links of respective specifications. Furthermore, each component is specified with corresponding pre- and post-conditions. However, the specification also includes a variety of event specifications.

In detail, the behavioral specification ("AppSpec"—as shown in Figure 2) of an application implementation ("AppImpl"—as shown in Figure 3) is described at the following two logical levels:

1. The *control level* describes the control structure of each of the component (e.g., sub-components, control flow and data flow links), which is:

   • Defined by the syntactic domain "StrModSeq", while the control flow can further be elaborated with syntactic domain "SplModSeq"

2. The *behavior level* describes the actual method's behavioral specification of each of component, which is defined by the syntactic domain "BehModSeq".

Furthermore, the registration of the *observations* is given by the syntactic domain "RegModSeq", at the top of the above domains. All four of the aforementioned domains are the top-level syntactic domains of the specification. Our specification is hierarchical, i.e., it specifies the components of the implementations as hierarchical modules. In the following, we discuss the syntax of *control* and *behavioral* elements of the specification using the specification of a temperature control as an example, shown in Figure 2.

1. The description of each component type consists of

   (a) Its interface, which is comprised of:

      • a list of inputs
      • a list of its outputs
      • a list of the resources it uses (e.g., files it reads, the code in memory that represents this component)
      • a list of sub-components required for the execution of the subject component
      • a list of events that represent entry into the component
      • a list of events that represent exit from the component
      • a list of events that are allowed to occur during any execution of this component
      • a set of conditional probabilities between the possible modes of the resources and the possible modes of the whole component
      • a list of known vulnerabilities occurred to the component

   (b) and a structural model that is a list of sub-components, some of which might be splits or joins of:

      • data-flows between linking ports of the sub-components (outputs of one to inputs of another)
      • control-flow links between cases of a branch and a component that will be enabled if that branch is taken.

The description of the component type is represented by the syntactical domain "StrMod", which is defined as follows:

StrMod ::− **define-ensemble** CompName
                      :entry-events :auto | (EvntSeq)
                      :exit-events (EvntSeq)
                      :allowable-events (EvntSeq)
                      :inputs (ObjNameSeq)
                      :outputs (ObjNameSeq)
                      :components (CompSeq)
                      :controlflows (CtrlFlowSeq)
                      :splits (SpltCFSeq)
                      :joins (JoinCFSeq)
                      :dataflows (DataFlowSeq)
                      :resources (ResSeq)

:**resource-mapping** (ResMapSeq)
:**model-mappings** (ModMapSeq)
:**vulnerabilities** (VulnrabltySeq)

Example 1. *For instance, a room temperature controller:* `control`, *periodically receives the current temperature value* (`sens-temp`) *from a sensor. The controller may also receive a user command* (`set-temp`) *to set (either increase or decrease) the current temperature of the room. Based on the received user command, the controller either raises the temperature through the sub-component* `temp-up` *or reduces the temperature through the sub-component* `temp-down`, *after computing the error* `compute-epsilon` *of the given command as shown in Figure 2. Furthermore, the controller issues a command as an output* (`com`), *which contains an updated temperature value. Figure 3 reflects the corresponding implementation parts of the controller.*

2. The behavioral specification of a component (a component type may have one normal behavioral specification and many abnormal behavioral specifications, each one representing some failure mode) consists of:

   - inputs and outputs
   - preconditions on the inputs (logical expressions involving one or more of the inputs)
   - postconditions (logical expressions involving one or more of the outputs and the inputs)
   - allowable events during the execution in this mode

   The behavioral specification of a component is represented by a corresponding syntactical domain "BehMod", as follows:

BehMod ::= **defbehavior-model** (CompName **normal | compromised**)
$\qquad\qquad$:**inputs** (ObjNameSeq$_1$)
$\qquad\qquad$:**outputs** (ObjNameSeq$_2$)
$\qquad\qquad$:**allowable-events** (EvntSeq)
$\qquad\qquad$:**prerequisites** (BehCondSeq$_1$)
$\qquad\qquad$:**post-conditions** (BehCondSeq$_2$)

Example 2. *For instance, in our temperature control example, the* `normal` *and* `compromised` *behavior of a controller component* `temp-up` *is modeled in Figure 2. The normal behavior of a temperature raising component* `temp-up` *describes the: (i) input condition* `prerequisites`, *i.e., the component receives a valid input* `new-temp`; *and (ii) output conditions* `post-conditions`, *i.e., the new computed temperature* `new-temp` *is equal to the sum of the current temperature* `old-temp` *and the computed* `delta` *(error). The computed temperature respects the temperature range (1-40). Similarly, the* `compromised` *behavior of the component illustrates the corresponding input and output conditions. Figure 3 reflects the corresponding implementation parts of the* `temp-up` *component.*

The complete syntactic details of the specification language are discussed in [5].

Based on the core idea of Lamport [6], we have defined the semantics of the specification as a state relationship to achieve the desired insight of the program's behavior. This does so by relating pre- and post-states [7]. For simplicity, we chose to discuss semantics of the behavioral domain "BehMod". The denotational semantics of the specification language is based on denotational algebras [8]. We define the result of semantic valuation function as a predicate. The behavioral relation (BehRel) is defined as a predicate over an environment, a pre-state, and a post-state. The corresponding relation is defined as:

$$\text{BehRel} := \mathbb{P}(\text{Environment} \times \text{State} \times \text{State}_\perp)$$

The valuation function for the abstract syntax domain "BehMod" values is defined as:

$$[\![\text{BehMod}]\!]: \text{Environment} \rightarrow \text{BehRel}$$

```
1  (define-component-type control
2   :entry-events (control)
3   :exit-events (control)
4   :allowable-events (update-temp compute-epsilon)
5   :inputs (set-temp sens-temp)
6   :outputs (com)
7
8   :components
9  ((temp-up :type temp-up :models (normal))
10  (temp-down :type temp-down :models (normal)) ... )
11
12   :dataflows
13  ((set-temp control set-temp temp-up)
14  (epsilon temp-up epsilon temp-down)...))
15
16
17  (define-component-type compute-delta
18    :Primitive t
19    :entry-events (compute-delta)
20    :exit-events (clip-delta)
21    :inputs (proportional integral derivative)
22    :outputs (delta)
23    :behavior-modes (normal)
24    )
25
26  (defbehavior-model (compute-delta normal)
27    :inputs (proportional integral derivative)
28    :outputs (delta)
29    :prerequisites ([data-type-of ?proportional number]
30    [data-type-of ?integral number]
31    [data-type-of ?derivative number])
32    :post-conditions ([data-type-of ?delta number])
33    )
34
35  (define-component-type temp-up
36    :entry-events (temp-up)
37    :exit-events (temp-up)
38    :inputs (delta old-temp time-step)
39    :outputs (new-temp)
40    :behavior-modes (normal compromised) )
41
42  (defbehavior-model (temp-up normal)
43    :inputs (delta old-temp time-step)
44    :outputs (new-temp)
45    :prerequisites ([data-type-of ?delta number])
46    :post-conditions ([and [and [data-type-of ?delta number]
47    [equal new-temp (+ ?old-temp ?delta)]]
48    [and [(<= new-temp 40)] [(>= new-temp 1)]]]))
49
50  (defbehavior-model (temp-up compromised)
51    :inputs (delta old-temp time-step)
52    :outputs (new-temp)
53    :prerequisites ()
54    :post-conditions (not [and [and [data-type-of ?delta number]
55    [equal new-temp (+ ?old-temp ?delta)]]
56    [and [(<= new-temp 40)] [(>= new-temp 1)]]]))
57
58
59  (define-component-type temp-down
60    :entry-events (temp-down)
61    :exit-events (temp-down)
62    :inputs (delta old-temp time-step)
63    :outputs (new-temp)
64    :behavior-modes (normal compromised) )
65
66  (defbehavior-model (temp-down normal)
67    :inputs (delta old-temp time-step)
68    :outputs (new-temp)
69    :prerequisites ([data-type-of ?delta number])
70    :post-conditions ([and [and [data-type-of ?delta number]
71    [equal new-temp (- ?old-temp ?delta)]]
72    [and [(<= new-temp 40)] [(>= new-temp 1)]]]))
73
74  (defbehavior-model (temp-down compromised)
75    :inputs (delta old-temp time-step)
76    :outputs (new-temp)
77    :prerequisites ()
78    :post-conditions (not [and [and [data-type-of ?delta number]
79    [equal new-temp (- ?old-temp ?delta)]]
80    [and [(<= new-temp 40)] [(>= new-temp 1)]]]))
81  ...
```

Figure 2. An example specification (AppSpec) of a temperature control.

Semantically, normal and compromised behavioral models result in modifying the corresponding elements of the environment value "Component" as defined below:

$[\![BehMod]\!](e)(e', s, s') \Leftrightarrow$

$\forall\, e_1 \in$ Environment, nseq $\in$ EvntNameSeq, eseq $\in$ ObsEvent*, inseq, outseq $\in$ Value*:

    $[\![ObjNameSeq_1]\!](e)(inState_\perp(s), inseq) \wedge [\![BehCondSeq_1]\!](e)\ (inState_\perp(s)) \wedge$

    $[\![EvntSeq]\!](e)\ (e_1, s, s', nseq, eseq)$

    $[\![ObjNameSeq_2]\!](e_1)(s', outseq) \wedge [\![BehCondSeq_2]\!](e_1)\ (s') \wedge$

    $\exists\, c \in$ Component: $[\![CompName]\!](e_1)(inValue(c)) \wedge$

    IF eqMode(inState$_\perp$(s'), "normal") THEN

        LET sbeh = c[1], nbeh = <inseq, outseq, s, s'>, cbeh = c[3] IN

                $e' = push(e_1, store(inState(s'))([\![CompName]\!](e_1)), c(sbeh, nbeh, cbeh, s, s'))$

        END

    ELSE

        LET sbeh = c[1], nbeh = c[2], cbeh = <inseq, outseq, s, s'> IN

                $e' = push(e_1, store(inState(s'))([\![CompName]\!](e_1)), c(sbeh, nbeh, cbeh, s, s'))$

        END

    END

In detail, if the semantics of the syntactic domain "BehMod" holds in a given environment e, resulting in environment e' and transforming a pre-state s into a corresponding post-state s', then:

- The inputs "ObjNameSeq$_1$" evaluate to a sequence of values *inseq* in a given environment e and a given state s, which satisfies the corresponding pre-conditions "BehCondSeq$_1$" in the same e and s.
- The allowable events happen and their evaluation results in new environment e_1 and a given post-state s' with some auxiliary sequences *nseq* and *eseq*.
- The outputs "ObjNameSeq$_2$" evaluates to a sequence of values *outseq* in an environment e_1 and given post-state s', which satisfies the corresponding post-conditions "BehCondSeq$_2$" in the same environment e_1. State s' and the given environment e' may be constructed such that:

  - If the post-state is "normal" then e' is an update to the normal behavior "nbeh" of the component "CompName" in environment e_1, otherwise
  - e' is an update to the compromised behavior "cbeh" of the component.

In the construction of the environment e', the rest of the semantics of the component do not change as represented in the corresponding LET-IN constructs.

The complete definitions of the auxiliary functions, predicates, and semantics are presented in [5].

3. An Execution Monitor of ARMET

In principle, an execution monitor interprets the event stream (traces of the execution of the target system, i.e., *observations*) against the system specification (the execution of the specification is also called *predictions*) by detecting inconsistencies between *observations* and the *predictions*, if there are any.

When the system implementation "AppImpl" (as shown in Figure 3) starts execution, an initial "startup" event is generated and dispatched to the top level component (module) of the system that transforms the execution state of the component into "running" mode. If there is a subnetwork of components, the component instantiates it and propagates the data along its data links by enabling the corresponding control links, if involved. When the data arrives on the input port of the component, the execution monitor checks if it is complete. If so, the execution monitor checks the preconditions of the component for the data and, if they succeed, it transforms the state of the component into "ready" mode. In the case that any of the preconditions fail, it enables the diagnosis engine.

After the startup of the implementation, described above, the execution monitor starts monitoring the arrival of every *observation* (runtime event) as follows:

1. If the event is a "method entry", the execution monitor checks if this is one of the "entry events" of the corresponding component in the "ready" state. If so, after receiving the data and when the respective preconditions are checked, if they succeed, the data is applied on the input port of the component and the mode of the execution state is changed to "running".
2. If the event is a "method exit", the execution monitor checks if this one of the "exit events" of the component is in the "running" state. If so, it changes its state into "completed" mode, collects the data from the output port of the component, and checks for corresponding postconditions. Should the checks fail, the execution monitor enables the diagnosis engine.
3. If the event is one of the "allowable events" of the component, it continues execution.
4. If the event is an unexpected event (i.e., it is neither an "entry event", an "exit event", nor in the "allowable events"), the execution monitor starts its diagnosis.

```
...
public class TemperatureControl{

    int old-temp = 0;
    int new-temp;

    public TemperatureControl(...){
        ...
    }

    public void set-temp(int temp){
        ...
        float delta = compute-delta(...);
        if(temp < old-temp)
            new-temp = temp-down(delta);
        else if(temp > old-temp)
            new-temp = temp-up(delta);
    }

    public double compute-delta(double proportional, double integral, double derivative){
        return proportional + integral + derivative;
    }

    public int temp-up(double delta){
        return old-temp + Math.round(delta);
    }

    public int temp-down(double delta){
        return old-temp - Math.round(delta);
    }

    ....
}
```

Figure 3. An example implementation (AppImpl) of a temperature control.

Based on the above behavioral description of the execution monitor, we have formalized the corresponding semantics of the execution monitor as follows:

$\forall$ app $\in$ AppImpl, sam $\in$ AppSpec, c $\in$ Component,
  s, s' $\in$ State$_c$, t, t' $\subset$ State$_s$, d, d' $\in$ Environment$_\partial$, e, e' $\in$ Environment, rte $\in$ RTEvent:
  [[sam]](d)(d', t, t') $\land$ [[app]](e)(e', s, s') $\land$ startup(s, app) $\land$ isTop(c, [[app]](e)(e', s, s')) $\land$
  setMode(s, "running") $\land$ arrives(rte, s) $\land$ equals(t, s) $\land$ equals(d, e)
  $\Rightarrow$
    $\forall$ p, p' $\in$ Environment*, m, n $\in$ State$^*_\perp$: equals(m(0), s) $\land$ equals(p(0), e)
    $\Rightarrow$
      $\exists$ k $\in$ $\mathbb{N}$, p, p' $\in$ Environment*, m, n $\in$ State$^*_\perp$:
      $\forall$ i $\in$ $\mathbb{N}_k$: monitors(i, rte, c, p, p', m, n) $\land$
        ((eqMode(n(k), "completed") $\land$ eqFlag(n(k), "normal") $\land$ equals(s', n(k)))
        $\lor$
        eqFlag(n(k), "compromised")))

$$\Rightarrow$$
$$\text{enableDiagnosis}(p'(k))(n(k), \text{inBool(true)}) \wedge \text{equals}(s', n(k)))$$

The semantics of recursive monitoring is determined by two sequences of states: pre and post, constructed from the pre-state of the monitor. Any *ith* iteration of the monitor transforms the $pre(i)$ state into the $post(i+1)$ state, from which the $pre(i+1)$ state is constructed. No event can be accepted in an *Error* state and the corresponding monitoring terminates when either the application has terminated with "normal" mode or when some misbehavior is detected, as indicated by the respective "compromised" state. This recursive idea of monitoring is formalized as a "monitors" predicate, as follows:

monitors $\subset \mathbb{N} \times$ **RTEvent** $\times$ **Component** $\times$ **Environment**$^* \times$ **Environment**$^* \times$ **State**$^* \times$ **State**$^*_\perp$
monitors(i, [[rte]], [[c]], e, e', s, s') $\Leftrightarrow$
(eqMode(s(i), "running") $\vee$ eqMode(s(i), "ready")) $\wedge$ [[c]](e(i))(e'(i), s(i), s'(i)) $\wedge$
  $\exists$ oe $\in$ ObEvent: equals(rte, store([[name(rte)]])(e(i))) $\wedge$
  IF entryEvent(oe, c) THEN
    data(c, s(i), s'(i)) $\wedge$
    (preconditions(c, e(i), e'(i), s(i), s'(i), "compromised") $\Rightarrow$ equals(s(i+1), s(i)) $\wedge$ equals(s'(i+1), s(i+1))
    $\wedge$ setFlag(inState(s'(i+1)), "compromised")) $\vee$ (preconditions(c, e(i), e'(i), s(i), s'(i), "normal")
    $\Rightarrow$ setMode(s(i), "running") $\wedge$
    LET cseq = components(c) IN
      equals(s(i+1), s'(i)) $\wedge$ equals(e(i+1), e'(i)) $\wedge$
      $\forall$ $c_1 \in$ cseq, $rte_1 \in$ RTEvent:
        arrives(rte_1, s(i+1)) $\wedge$ monitor(i+1, rte_1, c_1, e(i+1), e'(i+1), s(i+1), s'(i+1))
    END)
  ELSE IF exitEvent(oe, c) THEN
      data(c, s(i), s'(i)) $\wedge$ eqMode(inState(s'(i)), "completed") $\wedge$
      (postconditions(c, e(i), e'(i), s(i), s'(i), "compromised") $\Rightarrow$ equals(s(i+1), s(i)) $\wedge$ equals(s'(i+1), s(i+1))
      $\wedge$ setFlag(inState(s'(i+1)), "compromised")) $\vee$
      (postconditions(c, e(i), e'(i), s(i), s'(i), "normal") $\Rightarrow$ equals(s(i+1), s'(i)) $\wedge$ equals(e(i+1), e'(i)) $\wedge$
      setMode(inState(s'(i+1), "completed"))
  ELSE IF allowableEvent(oe, c) THEN equals(s(i+1), s'(i)) $\wedge$ equals(e(i+1), e'(i))
  ELSE equals(s(i+1), s(i)) $\wedge$ equals(s'(i+1), s(i+1)) $\wedge$ setFlag(inState(s'(i+1)), "compromised")
  END

The predicate "monitors" are defined as a relation on

- the number of observation i, with respect to the iteration of a component
- an observation (runtime event) *rte*
- the corresponding component c under observation
- a sequence of pre-environments e
- a sequence of post-environments e'
- a sequence of pre-states s
- a sequence of post-states s'

The predicate "monitors" are defined such that when an any arbitrary observation is made, if the current execution state $s(i)$ of component c is "ready" or "running", the behavior of component c has been evaluated, and there is a *prediction oe* that is semantically equal to an *observation rte*, any of the following can happen:

- The *prediction* or *observation* is an entry event of the component c and it waits until the complete data for the component c arrives. If this occurs then either:

- Preconditions of "normal" behavior of the component hold.If so, the subnetwork of the component is initiated and the components in the subnetwork are monitored iteratively with the corresponding arrival of the *observation*, or
- Preconditions of "compromised" behavior of the component hold. In this case, the state is marked as "compromised" and returns.

- The *observation* is an exit event and, after the completion of the data arrival, the postconditions hold and the resulting state is marked as "completed".
- The *observation* is an allowable event and just continues the execution.
- The *observation* is an unexpected event (or any of the above does not hold) and the state is marked as "compromised", and returns.

4. Proof of Behavioral Correctness

Based on the formalization of the denotational semantics of the specification language and the monitor, we have proved that the monitor is sound and complete, i.e., if the application implementation (AppImpl) is consistent with its specification (AppSpec), the security monitor will produce no false alarms (soundness) and the monitor will detect any deviation of the application execution from the behavior sanctioned by the specification language (completeness). In the following subsections, we articulate soundness and completeness statements and sketch their corresponding proofs.

4.1. Soundness

The intent of the soundness statement is to articulate whether the system's behavior is consistent with behavioral specification. Essentially, the goal is to show the absence of a false negative alarm such that whenever the security monitor alarms, there is a semantic inconsistency between the post-state of the program execution and the post-state of the specification execution. The soundness theorem is stated as follows:

Theorem 1 (Soundness). *The result of the security monitor is sound for any execution of the target system and its specification, iff, the specification is consistent with the program and the program executes in a safe pre-state and in an environment that is consistent with the environment of the specification, then*

- *for the pre-state of the program, there is an equivalent safe pre-state for which the specification can be executed and the monitor can be observed and*
- *if we execute the specification in an equivalent safe pre-state and observe the monitor at any arbitrary (combined) post-state, then*

   - *either there is no alarm, and then the post-state is safe and the program execution (post-state) is semantically consistent with the specification execution (post-state)*
   - *or there is an alarm, and then the post-state is compromised and the program execution (post-state) and the specification execution (post-state) are semantically inconsistent.*

Formally, the soundness theorem has the following signatures and definition.

Soundness $\subseteq \mathbb{P}(\text{AppImpl} \times \text{AppSpec} \times \text{Bool})$
Soundness$(\kappa, \omega, b) \Leftrightarrow$
$\forall e_s \in \text{Environment}_s, e_r, e_r' \in \text{Environment}_r, s, s' \in \text{State}_r$: consistent$(e_s, e_r) \land$ consistent$(\kappa, \omega) \land$
   $[\kappa](e_r)(e_r', s, s') \land$ eqMode$(s, \text{"normal"})$
$\Rightarrow$
   $\exists t, t' \in \text{State}_s, e_s' \in \text{Environment}_s$: equals$(s, t) \land [\omega](e_s)(e_s', t, t') \land$ monitor$(\kappa, \omega)(e_r; e_s)(s; t, s'; t') \land$
   $\forall t, t' \in \text{State}_s, e_s' \in \text{Environment}_s$: equals$(s, t) \land [\omega](e_s)(e_s', t, t') \land$ monitor$(\kappa, \omega)(e_r; e_s)(s; t, s'; t')$
   $\Rightarrow$
   LET $b = $ eqMode$(s', \text{"normal"})$ IN
      IF $b = $ True THEN equals(s', t') ELSE $\neg$ equals(s', t')

In detail, the soundness statement says that, if the following are satisfied:

1. a specification environment (e_s) is *consistent* with a run-time environment (e_r), and
2. a target system (κ) is *consistent* with its specification (ω), and
3. in a given run-time environment (e_r), execution of the system (κ) transforms the pre-state (s) into a post-state (s'), and
4. the pre-state (s) is safe, i.e., the state is in "normal" mode,

Then the following occurs:

- there exist the pre- and post-states (t and t', respectively) and the environment (e_s') of the specification execution such that in a given specification environment (e_s), the execution of the specification (ω) transforms the pre-state (t) into a post-state (t')
- the pre-states s and t are *equal* and *monitoring* of the system (κ) transforms the combined pre-state (s; t) into a combined post-state (s'; t')
- if both the following occur: in a given specification environment (e_s), execution of the specification (ω) transforms pre-state (t) into a post-state (t'); and the pre-states s and t are *equal* and *monitoring* of the system (κ) transforms the pre-state (s) into a post-state (s'), then either:

    - there is no alarm (b is True), the post-state s' of a program execution is safe, and the resulting states s' and t' are semantically *equal*, or
    - the security monitor alarms (b is False), the post-state s' of program execution is compromised, and the resulting states s' and t' are semantically not *equal*.

In the following section we present proof of the soundness statement.

Proof. The proof is essentially a structural induction on the elements of the specification (ω) of the system (κ). We have proved only the interesting case β of the specification to show that the proof works in principle. However, the proof of the remaining parts can easily be rehearsed following a similar approach.

The proof is based on certain lemmas, which are mainly about the relationships between different elements of the system and its specification (being at different levels of abstraction). These lemmas and relations can be proved based on the defined auxiliary functions and predicates that are based on the method suggested by Hoare [9]. The complete proof is presented in [10]. □

4.2. Completeness

The goal of the completeness statement is to show the absence of false positive alarms such that whenever there is a semantic inconsistency between the post-state of the program execution and the post-state of the specification execution, the security monitor alarms. The completeness theorem is stated as follows:

Theorem 2 (Completeness). *The result of the security monitor is complete for a given execution of the target system and its specification, iff, the specification is consistent with the program and the program executes in a safe pre-state and in an environment that is consistent with the environment of the specification, then*

- *for the pre-state of the program, there is an equivalent safe pre-state for which the specification can be executed and the monitor can be observed and*
- *if we execute the specification in an equivalent safe pre-state and observe the monitor at any arbitrary (combined) post-state, then*

    - *either the program execution (post-state) is semantically consistent with the specification execution (post-state), then there is no alarm and the program execution is safe*
    - *or the program execution (post-state) and the specification execution (post-state) are semantically inconsistent, then there is an alarm and the program execution has been compromised.*

Formally, the completeness theorem has the following signatures and definition.

Completeness $\subseteq \mathbb{P}($AppImpl $\times$ AppSpec $\times$ Bool$)$

Completeness$(\kappa, \omega, b) \Leftrightarrow$

$\forall\, e_s \in$ Environment$_s$, e_r, $e_r' \in$ Environment$_r$, s, $s' \in$ State$_r$: consistent$(e_s, e_r) \wedge$ consistent$(\kappa, \omega) \wedge$
$\quad [\![\kappa]\!](e_r)(e_r', s, s') \wedge$ eqMode$(s, $"normal"$)$

$\Rightarrow$

$\quad \exists\, t, t' \in$ State$_s$, $e_s' \in$ Environment$_s$: equals$(s, t) \wedge [\![\omega]\!](e_s)(e_s', t, t') \wedge$ monitor$(\kappa, \omega)(e_r;e_s)(s;t, s';t') \wedge$
$\quad \forall\, t, t' \in$ State$_s$, $e_s' \in$ Environment$_s$: equals$(s, t) \wedge [\![\omega]\!](e_s)(e_s', t, t') \wedge$ monitor$(\kappa, \omega)(e_r;e_s)(s;t, s';t')$

$\quad \Rightarrow$

$\quad\quad$ IF equals(s', t') THEN $b = $ True $\wedge b = $ eqMode$(s', $"normal"$)$
$\quad\quad$ ELSE $b = $ False $\wedge b = $ eqMode$(s', $"normal"$)$

In detail, the completeness statement says that, if the following are satisfied:

1. a specification environment (e_s) is *consistent* with a run-time environment (e_r), and
2. a target system (κ) is *consistent* with its specification (ω), and
3. in a given run-time environment (e_r), execution of the system (κ) transforms the pre-state (s) into a post-state (s'), and
4. the pre-state (s) is safe, i.e., the state is in "normal" mode,

Then the following occurs:

- there exist the pre- and post-states (t and t', respectively) and the environment (e_s') of specification execution such that, in a given specification environment (e_s), execution of the specification (ω) transforms the pre-state (t) into a post-state (t')
- the pre-states s and t are *equal* and *monitor*ing of the system (κ) transforms the combined pre-state (s; t) into a combined post-state (s'; t')
- if both: in a given specification environment (e_s), the execution of the specification (ω) transforms the pre-state (t) into a post-state (t'); and the pre-states s and t are *equal* and *monitor*ing of the system (κ) transforms the pre-state (s) into a post-state (s'), then either

  - the resulting two post-states s' and t' are semantically *equal* and there is no alarm, or
  - the resulting two post-states s' and t' are semantically not *equal* and the security monitor alarms.

Proof. The proof of completeness is very similar to the proof of soundness. The complete proof is presented in [10]. □

5. Conclusions

We have presented a formalization of the semantics of the specification language and monitor of the cognitive middleware ARMET. In order to assure the continuous operation of ICS applications and to meet the real-time requirements of ICS, we have proved that our run-time security monitor produces no false alarm and always detects behavioral deviation of the ICS application. We plan to integrate our run-time security monitor with a security-by-design component to ensure comprehensive security solution for ICS applications.

Acknowledgments: The authors thank the anonymous reviewers on the earlier version of this work.

Author Contributions: All authors of the paper have contributed to the presented results. The ARMET prototype is based on the AWDRAT software developed by Howard Shrobe.

Conflicts of Interest: The authors declare no conflict of interest.

References

1. Khan, M.T.; Serpanos, D.; Shrobe, H. A Rigorous and Efficient Run-time Security Monitor for Real-time Critical Embedded System Applications. In Proceedings of the 2016 IEEE 3rd World Forum on Internet of Things (WF-IoT), Reston, VA, USA, 12–14 December 2016; pp. 100–105.
2. Shrobe, H.; Laddaga, R.; Balzer, B.; Goldman, N.; Wile, D.; Tallis, M.; Hollebeek, T.; Egyed, A. AWDRAT: A Cognitive Middleware System for Information Survivability. In Proceedings of the IAAI'06, 18th Conference on Innovative Applications of Artificial Intelligence, Boston, MA, USA, 16–20 July 2006; AAAI Press: San Francisco, CA, USA, 2006; pp. 1836–1843.
3. Shrobe, H.E. *Dependency Directed Reasoning for Complex Program Understanding*; Technical Report; Massachusetts Institute of Technology: Cambridge, MA, USA, 1979.
4. Lynx Software Technologies. LynxOS. Available online: http://www.lynx.com/industry-solutions/industrial-control/ (accessed on 20 July 2017).
5. Khan, M.T.; Serpanos, D.; Shrobe, H. *On the Formal Semantics of the Cognitive Middleware AWDRAT*; Technical Report MIT-CSAIL-TR-2015-007; Computer Science and Artificial Intelligence Laboratory, MIT: Cambridge, MA, USA, 2015.
6. Lamport, L. The temporal logic of actions. *ACM Trans. Program. Lang. Syst.* **1994**, *16*, 872–923.
7. Khan, M.T.; Schreiner, W. Towards the Formal Specification and Verification of Maple Programs. In *Intelligent Computer Mathematics*; Jeuring, J., Campbell, J.A., Carette, J., Reis, G.D., Sojka, P., Wenzel, M., Sorge, V., Eds.; Springer: Berlin, Germany, 2012; pp. 231–247.
8. Schmidt, D.A. *Denotational Semantics: A Methodology for Language Development*; William, C., Ed.; Brown Publishers: Dubuque, IA, USA, 1986.
9. Hoare, C.A.R. Proof of correctness of data representations. *Acta Inform.* **1972**, *1*, 271–281.
10. Khan, M.T.; Serpanos, D.; Shrobe, H. *Sound and Complete Runtime Security Monitor for Application Software*; Technical Report MIT-CSAIL-TR-2016-017; Computer Science and Artificial Intelligence Laboratory, MIT: Cambridge, MA, USA, 2016.

 *electronics*

Article

μRTZVisor: A Secure and Safe Real-Time Hypervisor

José Martins [†], João Alves [†], Jorge Cabral, Adriano Tavares and Sandro Pinto *

Centro Algoritmi, Universidade do Minho, 4800-058 Guimarães, Portugal; jose.martins@dei.uminho.pt (J.M.); joao.alves@dei.uminho.pt (J.A.); jcabral@dei.uminho.pt (J.C.); atavares@dei.uminho.pt (A.T.)
* Correspondence: sandro.pinto@dei.uminho.pt; Tel.: +351-253-510-180
† These authors contributed equally to this work.

Received: 29 September 2017; Accepted: 24 October 2017; Published: 30 October 2017

Abstract: Virtualization has been deployed as a key enabling technology for coping with the ever growing complexity and heterogeneity of modern computing systems. However, on its own, classical virtualization is a poor match for modern endpoint embedded system requirements such as safety, security and real-time, which are our main target. Microkernel-based approaches to virtualization have been shown to bridge the gap between traditional and embedded virtualization. This notwithstanding, existent microkernel-based solutions follow a highly para-virtualized approach, which inherently requires a significant software engineering effort to adapt guest operating systems (OSes) to run as userland components. In this paper, we present μRTZVisor as a new TrustZone-assisted hypervisor that distinguishes itself from state-of-the-art TrustZone solutions by implementing a microkernel-like architecture while following an object-oriented approach. Contrarily to existing microkernel-based solutions, μRTZVisor is able to run nearly unmodified guest OSes, while, contrarily to existing TrustZone-assisted solutions, it provides a high degree of functionality and configurability, placing strong emphasis on the real-time support. Our hypervisor was deployed and evaluated on a Xilinx Zynq-based platform. Experiments demonstrate that the hypervisor presents a small trusted computing base size (approximately 60KB), and a performance overhead of less than 2% for a 10 ms guest-switching rate.

Keywords: virtualization; hypervisor; TrustZone; microkernel; security; safety; real-time; Arm

---

1. Introduction

Embedded systems were, for a long time, single-purpose and closed systems, characterized by hardware resource constraints and real-time requirements. Nowadays, their functionality is ever-growing, coupled with an increasing complexity that is associated with a higher number of bugs and vulnerabilities [1,2]. Moreover, the pervasive connectivity of these devices in the modern Internet of Things (IoT) era significantly increases their attack surface [3,4]. Due to their myriad of applications and domains, ranging from consumer electronics to aerospace control systems, there is an increasing reliance on embedded devices that often have access to sensitive data and perform safety-critical operations [5–7]. Thus, two of the main challenges faced by modern embedded systems are those of security and safety. Virtualization emerged as a natural solution, due to the isolation and fault-containment it provides, by encapsulating each embedded subsystem in its own virtual machine (VM). This technology also allows for different applications to be consolidated into one single hardware platform, thus reducing size, weight, power and cost (SWaP-C) budgets, at the same time providing an heterogeneous operating system (OS) environment fulfilling the need for high-level programming application programming interfaces (API) coexisting alongside real-time functionality and even legacy software [5,8,9].

Despite the differences among several embedded industries, all share an increasing interest in exploring virtualization mainly for isolation and system consolidation. For example, in consumer electronics, due to current smartphone ubiquity and IoT proliferation, virtualization is being used to isolate the network stack from other commodity applications or GUI software, as well as for segregating sensitive data (e.g., companies relying on mobile phones isolate enterprise from personal data) [5,10,11]. In modern industrial control systems (ICSs), there is an increasing trend for integrating information technology (IT) with the operation technology (OT). In this context, ICSs need to guarantee functionality isolation, while protecting their integrity against unauthorized modification and restricting access to production related data [12,13]. The aerospace and the automotive industries are also dependent on virtualization solutions due to the amount of needed control units, which, prior to virtualization application, would require a set of dedicated microcontrollers, and which are currently being consolidated into one single platform. In the case of aerospace systems, virtualization allows this consolidation to guarantee the Time and Space Partitioning (TSP) required for the reference Integrated Modular Avionics (IMA) architectures [14]. As for the automotive industry, it also allows for safe coexistence of safety-critical subsystems with real-time requirements and untrusted ones such as infotainment applications [6]. Finally, in the development of medical devices, which are becoming increasingly miniaturized, virtualization is being applied to consolidate their subsystems and isolate their critical life-supporting functionality from communication or interface software used for their control and configuration, many times operated by the patient himself. These systems are often composed of large software stacks and heavy OSes containing hidden bugs and that, therefore, cannot be trusted [5].

Virtualization software, dubbed virtual machine monitor (VMM) or hypervisor, must be carefully designed and implemented since it constitutes a single point of failure for the whole system [15]. Hence, this software, the only one with maximum level of privilege and full access to all system resources, i.e., the trusted computing base (TCB), should be as small and simple as possible [16,17]. In addition, virtualization on its own is a poor match for modern embedded systems, given that for their modular and inter-cooperative nature, the strict confinement of subsystems interferes with some fundamental requirements [1,8,9]. First, the decentralized, two-level hierarchical scheduling inherent to virtualization interferes with its real-time capabilities [18,19]. In addition, the strongly isolated virtual machine model prevents embedded subsystems to communicate in an efficient manner.

Microkernel-based approaches to virtualization have been shown to bridge the gap between traditional virtualization and current embedded system requirements. Contrarily to monolithic hypervisors, which implement the VM abstraction and other non-essential infrastructure such as virtual networks in a single, privileged address space, microkernels implement a minimal TCB by only providing essential policy-void mechanisms such as thread, memory and communication management, leaving all remaining functionality to be implemented in userland components [20,21]. By employing the principles of minimality and of the least authority, this architecture has proven to be inherently more secure [22,23] and a great foundation to manage the increasing complexity of embedded systems [24,25]. However, existent microkernel-based solutions follow a highly para-virtualized approach. OSes are hosted as user-level servers providing the original OS interfaces and functionality through remote procedure calls (RPC), and, thus, these must be heavily modified to run over the microkernel [26,27]. Microkernel-based systems also seem to be well-suited for real-time environments, and, because of their multi-server nature, the existence of low-overhead inter-partition communication (IPC) is a given [28,29].

Aware of the high overhead incurred by trap-and-emulate and para-virtualization approaches, some processor design and manufacturing companies have introduced support for hardware virtualization to their architectures such as Intel's VT-x (Intel, Santa Clara, CA, USA) [30,31], Arm's VE (ARM Holdings, Cambridge, England, United Kingdom) [32,33] or Imagination's MIPS VZ (Imagination Technologies, Kings Langley, Hertfordshire, United Kingdom) [34]. These provide a number of features to ease the virtualization effort and minimize hypervisor size such as the introduction of higher privilege modes, configuration hardware replication and multiplexing, two-level address translation or virtual interrupt support. Arm TrustZone hardware extensions, although being a security oriented technology, provide similar features to those aforementioned. However, they do not provide two-level address translation but only memory segmentation support. Hence, although guests can run nearly unmodified, they need to be compiled and cooperate to execute in the confinement of their attributed segments. Despite this drawback and given that this segmented memory model is likely to suffice for embedded use-cases (small, fixed number of VMs), these extensions have been explored to enable embedded virtualization, also driven by the fact that its deployment is widely spread in low-end and mid-range microprocessors used in embedded devices [35–40]. Existing TrustZone virtualization solutions show little functionality and design flexibility, consistently employing a monolithic structure. Despite taking advantage of security extensions, TrustZone hypervisors focus exclusively on virtualization features leaving aside important security aspects.

Under the light of the above arguments, in this work, we present μRTZVisor as a new TrustZone-assisted hypervisor that distinguishes itself from existing TrustZone-assisted virtualization solutions by implementing a microkernel-like architecture while following an object-oriented approach. The μRTZVisor's security-oriented architecture provides a high degree of functionality and configuration flexibility. It also places strong emphasis on real-time support while preserving the close to full-virtualized environment typical of TrustZone hypervisors, which minimizes the engineering effort needed to support unmodified guest OSes. With μRTZVisor, we make the following contributions:

1. A heterogeneous execution environment supporting coarse-grained partitions, destined to run guest OSes on the non-secure world, while providing user-level finer-grained partitions on the secure side, used for implementing encapsulated kernel extensions.
2. A real-time, priority and time-partition based scheduler enhanced by a timeslice donation scheme which guarantees processing bandwidth for each partition. This enables the co-existence of non real-time and real-time partitions with low interrupt latencies.
3. Secure IPC mechanisms wrapped by a capability-based access control system and tightly coupled with the real-time scheduler and memory management facilities, enabling fast and efficient partition interactions.
4. Insight on the capabilities and shortcomings of TrustZone-based virtualization, since this design orbited around the alleviation of some of the deficiencies of TrustZone support for virtualization.

The remainder of this paper is structured as follows: Section 2 provides some background information by overviewing Arm TrustZone technology and RTZVisor. Section 3 describes and explains all implementation details behind the development of μRTZVisor: architecture, secure boot, partition and memory manager, capability manager, device and interrupt manager, IPC manager, and scheduler. Section 4 evaluates the hypervisor in terms of memory footprint (TCB size), context-switch overhead, interrupt latency, and IPC overhead. Section 5 points and describes related work in the field. Finally, Section 6 concludes the paper and highlights future research directions.

2. Background

This section starts by detailing the Arm TrustZone security extensions, the cornerstone technology on which μRTZVisor relies. It also describes RTZVisor, the previous version of the hypervisor on which our implementation is built and which persists as the kernel's foundation for essential virtualization mechanisms.

2.1. Arm TrustZone

TrustZone technology is a set of hardware security extensions, which have been available on Arm Cortex-A series processors for several years [41] and has recently been extended to cover the new generation Cortex-M processor family. TrustZone for Armv8-M has the same high-level features as TrustZone for applications processors, but it is different in the sense that the design is optimized for microcontrollers and low-power applications. In the remainder of this section, when describing TrustZone, the focus will be on the specificities of this technology for Cortex-A processors (Figure 1a).

The TrustZone hardware architecture virtualizes a physical core as two virtual cores, providing two completely separated execution environments: the *secure* and the *non-secure* worlds. A new 33rd processor bit, the *Non-Secure (NS)* bit, indicates in which world the processor is currently executing. To preserve the processor state during the world switch, TrustZone adds an extra processor mode: the monitor mode. The monitor mode is completely different from other modes because, when the processor runs in this privileged mode, the state is always considered secure, independently of the NS bit state. Software stacks in the two worlds can be bridged via a new privileged instruction-*Secure Monitor Call (SMC)*. The monitor mode can also be entered by configuring it to handle interrupts and exceptions in the secure side. To ensure a strong isolation between secure and non-secure states, some special registers are banked, while others are either totally unavailable for the non-secure side.

The TrustZone Address Space Controller (TZASC) and the TrustZone Memory Adapter (TZMA) extend TrustZone security to the memory infrastructure. TZASC can partition the dynamic random-access memory (DRAM) into different secure and non-secure memory regions, by using a programming interface which is only accessible from the secure side. By design, secure world applications can access normal world memory, but the reverse is not possible. TZMA provides similar functionality but for off-chip read-only memory (ROM) or static random-access memory (SRAM). Both the TZASC and TZMA are optional and implementation-specific components on the TrustZone specification. In addition, the granularity of memory regions depends on the system on chip (SoC). The TrustZone-aware memory management unit (MMU) provides two distinct MMU interfaces, enabling each world to have a local set of virtual-to-physical memory address translation tables. The isolation is still available at the cache-level because processor's caches have been extended with an additional tag that signals in which state the processor accessed the memory.

System devices can be dynamically configured as secure or non-secure through the TrustZone Protection Controller (TZPC). The TZPC is also an optional and implementation-specific component on the TrustZone specification. To support the robust management of secure and non-secure interrupts, the Generic Interrupt Controller (GIC) provides both secure and non-secure prioritized interrupt sources. The interrupt controller supports interrupt prioritization, allowing the configuration of secure interrupts with a higher priority than the non-secure interrupts. Such configurability prevents non-secure software from performing a denial-of-service (DOS) attack against the secure side. The GIC also supports several interrupt models, allowing for the configuration of interrupt requests(IRQs) and fast interrupt requests (FIQs) as secure or non-secure interrupt sources. The suggested model by Arm proposes the use of IRQs as non-secure world interrupt sources, and FIQs as secure interrupt sources.

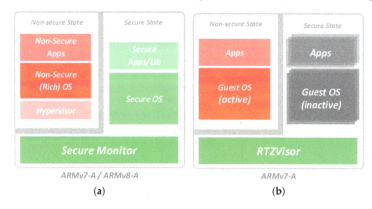

Figure 1. Arm TrustZone: generic architecture and RTZVisor system architecture. (**a**) Arm TrustZone architecture; (**b**) RTZVisor system architecture.

2.2. RTZVisor

RTZVisor [37], the Real-Time TrustZone-assisted Hypervisor, is a bare-metal embedded hypervisor that relies on TrustZone hardware to provide the foundation to implement strong spatial and temporal isolation between multiple guest OSes. RTZVisor is implemented in the C language and follows a monolithic architecture (Figure 1). All hypervisor components, drivers and other critical parts of the virtualization infrastructure run in the most privileged processor mode, i.e., the monitor mode. The hypervisor follows a simple and static implementation approach. All data structures and hardware resources are predefined and configured at design time, avoiding the use of language dynamic features.

Guest OSes are multiplexed on the non-secure world side; this requires careful handling of shared hardware resources, such as processor registers, memory, caches, MMU, devices, and interrupts. Processor registers are preserved in a specific virtual machine control block (VMCB). This virtual processor state includes the core registers for all processor modes, the CP15 registers and some registers of the GIC. RTZVisor offers as many vCPUs as the hardware provides, but only a one-to-one mapping between vCPU, guest and real CPU is supported. RTZVisor only offers the ability to create non-secure guest partitions, and no means of executing software in secure supervisor or user modes.

The strong spatial isolation is ensured through the TZASC, by dynamically changing the security state of the memory segments. Only the guest partition currently running in the non-secure side has its own memory segment configured as non-secure, while the remaining memory is configured as secure. The granularity of the memory segments, which is platform-dependent, limits the number of supported virtual machines (VMs). Moreover, since TrustZone-enabled processors only provide MMU support for single-level address translation, it means that guests have to know the physical memory segment they can use in the system, requiring relocation and consequent recompilation of the guest OS. Temporal isolation is achieved through a cyclic scheduling policy, ensuring that one guest partition cannot use the processor for longer than its defined CPU quantum. The time of each slot can be different for each guest, depending on its criticality classification, and is configured at design time. Time management is achieved by implementing two levels of timing: there are timing units for managing the hypervisor's time, as well as for managing the partitions' time. Whenever the active guest is executing, the timers belonging to the guest are directly managed and updated by the guest on each interrupt. For inactive guests, the hypervisor implements a virtual tickless timekeeping mechanism, which ensures that when a guest is rescheduled, its internal clocks and related data structures are updated with the time elapsed since its previous execution.

RTZVisor's main goal was to prove it was possible to run multiple guest OSes concurrently, completely isolated from each other, on TrustZone-enabled Arm processors without VE support. Despite achieving such a goal, RTZVisor still presented some limitations and open-issues. A list of the main identified limitations follow:

- Hypervisors are not magic bullets and they are also prone to incorrect expectations in terms of security. Guaranteeing a system is secure just by relying on a virtualization layer is not enough. These incorrect expectations probably come from the fact that a hypervisor provides separation and isolation, which positively impacts security. The problem is that security is much more than separation. Security starts from the onset, and hypervisors must be complemented with other security-oriented technologies for guaranteeing a complete chain of trust. The secure boot process is responsible for establishing a chain of trust that authenticates and validates all levels of secure software running on the device. In this sense, the integrity of the hypervisor at boot time is guaranteed.
- RTZVisor does not implement and enforce any existing coding standards. The use of coding standards is becoming imperative in modern security and safety-critical systems to reduce the number of programming errors and achieve certification.
- Although RTZVisor provides real-time support mainly by implementing efficient time services, these are still guest OS dependent and limited to a cyclic scheduling algorithm. The implementation does not allow for event-driven guests to preempt others, resulting in high interrupt latencies.

- The nature of embedded systems requires communication and interaction among the various subsystems. RTZVisor fails in this aspect by not implementing any kind of IPC facilities. All of its guests are completely isolated and encapsulated, having no mechanism to cooperate.
- Finally, and taking into account the previous point, RTZVisor provides no mechanisms for device sharing. Some kind of communication is needed for guests to synchronize, when accessing the same peripheral.

3. μRTZVisor

μRTZVisor is based on a refactoring of RTZVisor, designed to achieve a higher degree of safety and security. In this spirit, we start by anchoring our development process in a set of measures that target security from the onset. First, we made a complete refactoring of the original code from C to C++. The use of an object-oriented language promotes a higher degree of structure, modularity and clarity on the implementation itself, while leveraging separation of concerns and minimizing code entanglement. Kernel modules have bounded responsibilities and only interact through well-defined interfaces, each maintaining its internal state while sharing the control structure of each partition. However, we apply only a subset of C++ suitable to embedded systems, removing features such as multiple inheritance, exception handling or RTTI (Run-Time Type Information) which are error prone, difficult to understand and maintain, as well as unpredictable and inefficient from a memory footprint and execution perspective. In addition to the fact that C++ already provides stronger type checking and linkage, we reinforce its adoption by applying the MISRA (Motor Industry Software Reliability Association) C++ coding guidelines. Due to the various pitfalls of the C++ language, which make it ill-advised for developing critical systems, the main objective of the MISRA C++ guidelines is to define a safer subset of the C++ language suitable for use in safety related embedded systems. These guidelines were enforced by the use of a static code analyzer implementing the standard.

The main feature of μRTZVisor is its microkernel-like architecture, depicted in Figure 2. Nevertheless, we don't strive to implement traditional microkernel virtualization, which, given its para-virtualization nature, imposes heavy guest source-code modification. We aim at gathering those ideas that benefit security and design flexibility, while persevering the capability to run nearly unmodified guest OSes. Since TrustZone-enabled processors already provide two virtual CPUs, by providing a secure and non-secure view of the processor, and extends this partitioning to other resources such interrupts and devices, guests can make full use of all originally intended privileged levels, being allowed to directly configure assigned system resources, manage their own page tables and directly receive their assigned interrupts. However, the lack of a two-level address translation mechanism imposes a segmented memory model for VMs. Hence, guest OSes need to be compiled and cooperate to execute in the confinement of their assigned segments. This issue is augmented by the fact that segments provided by the TZASC are typically large (in the order of several MB) and must be consecutively allocated to guests, which leads to high levels of internal fragmentation. In addition, the maximum possible number of concurrent guests is limited by the total number of available TZASC segments, which varies according to the platform's specific implementation. Nevertheless, however small the number of segments, this segmented memory model is likely to suffice for embedded use-cases that usually require a small, fixed number of VMs according to deployed functionalities.

Besides this, guest OSes only need to be modified if they are required to use auxiliary services or shared resources that rely on the kernel's IPC facilities. For this, typically used commodity operating systems, such as Linux, may simply add kernel module drivers to expose these mechanisms to user applications. Multi-guest support is achieved by multiplexing them on the non-secure side of the processor, i.e., by dynamically configuring memory segments, devices or interrupts of the active partition as non-secure, while inactive partition resources are set as secure and by saving and restoring the context of CPU, co-processor and system control registers, which are banked between the two worlds. An active guest that attempts to access secure resources triggers an abort exception directly to the hypervisor. However, these exceptions may be imprecise as in the case of those triggered by an unallowed device access. Additionally, the existence of banked registers or their silently failing access, as is the case of secure interrupts in the GIC, coupled with the fact that the execution of privileged instructions cannot be detected by the monitor, makes it such that classical techniques such as trap-and-emulate cannot be used to further enhance a full-virtualization environment. At the moment, when one of the aforementioned exceptions is triggered, guests are halted or blindly rebooted. In addition, given that guests share cache and TLB (Translation Lookaside Buffer) infrastructures, these must be flushed when a new guest becomes active. Otherwise, the entries of the previous guest, which are marked as non-secure, could be accessed without restriction by the incoming one.

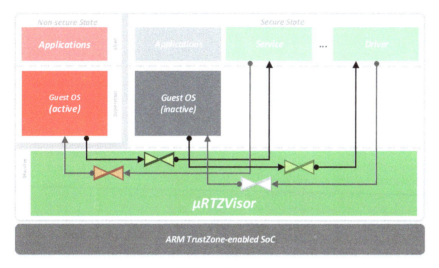

Figure 2. μRTZVisor architectural overview.

μRTZVisor privilege code runs in monitor mode, the most privileged level in TrustZone-enabled processors, having complete access and configuration rights over all system resources. This layer strives to be a minimal TCB, implementing only essential infrastructure to provide the virtual machine abstraction, spatial and temporal partitioning, and basic services such as controlled communication channels. The kernel's design aims for generality and flexibility so that new functionality can be added in a secure manner. For this, it provides a heterogeneous partition environment. As described above, coarse-grained partitions based on the memory segmentation model are used to run guest OSes. In addition, partitions running in secure user mode are implemented by managing page tables used by the MMU's secure interface, which allows for a greater degree of control over their address spaces. Secure user mode partitions are used to implement extra functionality, which would typically execute in kernel mode in a monolithic system. They act as server tasks that can be accessed through RPC operations sitting on the IPC and scheduling infrastructure. For example, shared device drivers or virtual network infrastructures can be encapsulated in these partitions. Herein lies the main inspiration from microkernel ideas. Non-essential services are encapsulated in these partitions, preventing fault-propagation to other components. Hence, they can be untrusted and developed by third-parties, incorporating only the TCB of other partitions that depend on them. Although these kind of services could be implemented in VMs running in the non-secure world, rendering worthless the extra core complexity added to the kernel, implementing them as secure world tasks provides several benefits. First, running them on a secure virtual address space eliminates the need for the relocation and recompilation and reduces the fragmentation inherent to the segmented memory model. This facilitates service addition, removal or swapping according to guests' needs and overall system requirements. At the same time, it enables finer-grained functionality fault-encapsulation. Finally, both the hypervisor and secure tasks always run with caches enabled, but, since caches are TrustZone-aware, there is no need to flush them when switching from a guest partition to a secure world task due to a service request via RPC, which significantly improves performance.

A crucial design decision relates to the fact that partitions are allocated statically, at compile-time. Given the static nature of typical embedded systems, there is no need for partitions to create other partitions or to possess parent-child relations and some kind of control over one another. This greatly simplifies the implementation of partition management, communication channel and resource distribution, which are defined and fixed according to the system design and configuration. This idea is further advanced in the next few paragraphs.

To achieve robust security, fault-encapsulation is not enough and the principle of the least authority must be thoroughly enforced. This is done at a first level by employing the aforementioned hardware mechanisms provided both by typical hardware infrastructure (virtual address translation or multiple privilege levels) and the TrustZone features that allow control over memory segments, devices and interrupts. Those are complemented by a capability-based access control mechanism. A capability represents a kernel object or a hardware resource and a set of rights over it. For example, a partition holding a capability for a peripheral can access it according to the set read/write permissions. This is essential so that secure tasks can configure and interact with their assigned hardware resources in an indirect manner, by issuing hypercalls to the kernel. Guest OSes may own capabilities but do not necessarily have to use them, unless they represent abstract objects such as communications endpoints, or the guest needs to cooperate with the kernel regarding some particular resource it owns. In this way, all the interactions with the kernel, i.e., hypercalls, become an invocation of an object operation through a capability. This makes the referencing of a resource by partitions conceptually impossible if they do not own a capability for it. Given that the use of capabilities provides fine-grained control over resource distribution, system configuration almost fully reduces to capability assignment, which shows to be a simple, yet flexible mechanism.

Built upon the capability system, this architecture provides a set of versatile inter-partition communication primitives, the crucial aspect of the microkernel philosophy. These are based on the notion of a port, constituting an endpoint to and from which partitions read and write messages. Given that these operations are performed using port capabilities, this enables system designers to accurately specify the existing communication channels. In addition, the notion of reply capabilities, i.e., port capabilities with only the send rights set, which can only be used once, and that are dynamically assigned between partitions through IPC, is leveraged to securely perform client-server type communications, since they remove the need to grant servers full-time access to client ports. Port operations can be characterized as synchronous or asynchronous, which trade-off security and performance. Asynchronous communication is safer since it doesn't require a partition to block, but entails some performance burden. In opposition, synchronous communication is more dangerous, since partitions may block indefinitely, but allows partitions to communicate faster, by integration with scheduler functionalities for efficient RPC communication. Aiming at providing the maximum design flexibility, our architecture provides both kinds of communication.

This architecture categorically differs from classical TrustZone software architectures, which typically feature a small OS running in secure supervisor mode that manages secure tasks providing services to non-secure partitions and that only execute when triggered by non-secure requests or interrupts. This service provision by means of RPC overlaps with our approach, but it focuses only on providing a Trusted-Execution Environment (TEE) and not on a flexible virtualized real-time environment. No such OS exists following this approach, since the hypervisor directly manages these tasks, leaving the secure supervisor mode vacant. This partial flattening of the scheduling infrastructure allows for the direct switch between guest client and server partitions, reducing overhead, and for secure tasks to be scheduled in their own right to perform background computations. At the same time, given that the same API is provided to both client and server partitions, it homogenizes the semantics of communication primitives and enables simple applications that show no need for a complete OS stack or large memory requirements to execute directly as secure tasks. In addition, in some microkernel-based virtualization implementations, the VM abstraction is provided by user-level components [31], which, in our system, would be equivalent to the secure tasks. This encompasses high-levels of IPC traffic between the VM and the guest OS and a higher number of context-switches. Given the lightweight nature of the VM provided by our system, this abstraction directly provided at the kernel level, which, despite slightly increasing TCB complexity, significantly reduces such overhead.

Besides security, the architecture places strong emphasis on the real-time guarantees provided by the hypervisor. Inspired by ideas proposed in [8], the real-time scheduler structure is based on the notion of time domains that execute in a round-robin fashion and to which partitions are statically assigned. This model guarantees an execution budget for each domain which is replenished after a complete cycle of all domains. Independently of their domain, higher priority partitions may preempt the currently executing one, so that event-driven partitions can handle events such as interrupts as quickly as possible. However, the budget allocated to these partitions must be chosen with care according to the frequency of the events, to not be exhausted, delaying the handling of the event until the next cycle. We enhance this algorithm with a time-slice donation scheme [42] in which a client partition may explicitly donate its domain's bandwidth to the target server until it responds, following an RPC pattern. In doing so, we allow for the co-existence of non-real time and real-time partitions, both time and event-driven, while providing fast and efficient communication interactions between them. All related parameters such as the number of domains, their budgets, partition allocation and their priorities are assigned at design time, providing once again a highly flexible configuration mechanism.

For the kernel's internal structure, we opted for a non-preemptable, event-driven execution model. This means that we use a single kernel stack across execution contexts, which completely unwinds when leaving the kernel, and that, when inside the kernel, interrupts are always disabled. Although this design may increase interrupt and preemption latencies, which affect determinism by increasing jitter, the additional needed complexity to make the kernel fully preemptable or support preemption points or continuations would significantly increase the system's TCB. Moreover, although the great majority of hypercalls and context switch operations show a short constant execution time, others display a linear time complexity according to hypercall arguments and the current state of time-slice donation, which may aggravate the aforementioned issues. We also maintain memory mappings enabled during hypercall servicing. This precludes the need to perform manual address translation on service call parameters located in partition address spaces, but it requires further cache maintenance to guarantee memory coherency in the eventuality that guests are running with caches disabled.

Finally, it is worth mentioning that the design and implementation of the μRTZVisor was tailored for a Zynq-7000 SoC and is heavily dependent on the implementation of TrustZone features on this platform. Although the Zynq provides a dual-core Arm Cortex-A9, the hypervisor only supports a single-core configuration. Support for other TrustZone-enabled platforms and multicore configurations will be explored in the near future.

3.1. Secure Boot

Apart from the system software architecture, security starts by ensuring a complete secure boot process. The secure boot process is made of a set of steps that are responsible for establishing a complete chain of trust. This set of steps validates, at several stages, the authenticity and integrity of the software that is supposed to run on the device. For guaranteeing a complete chain of trust, hardware trust anchors must exist. Regarding our current target platform, a number of secure storage facilities were identified, which include volatile and non-volatile memories. Off-chip memories should only be used to store secure encrypted images (boot time), or non-trusted components such as guest partitions (runtime).

The complete secure boot sequence is summarized and depicted in Figure 3. After the power-on and reset sequences have been completed, the code on an on-chip ROM begins to execute. This ROM is the only component that cannot be modified, updated or even replaced by simple reprogramming attacks, acting as the root of trust of the system. It starts the whole security chain by ensuring authentication and decryption of the first-stage bootloader (FSBL) image. The decrypted FSBL is then loaded into a secure on-chip memory (OCM). Once the FSBL starts to run, it is then responsible for the authentication, decryption and loading of the complete system image. This image contains the critical code of the μRTZVisor and secure tasks, as well as the guest OSes images. The binary images of the guest OSes are individually compiled for the specific memory segments they should run from, and then attached to the final system image through the use of specific assembly directives. Initially, they are positioned in consecutive (secure) memory addresses, and, later, the hypervisor is the one responsible for copying the individual guest images to the assigned memory segment they should run from. Therefore, at this stage, the system image is attested as a whole, and not individually. In the meantime, if any of the steps on the authentication and decryption of the FSBL or the system image fails, the CPU is set into a secure lockdown state.

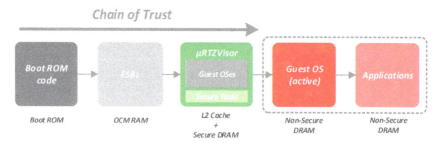

Figure 3. μRTZVisor: secure boot process.

Once the system image has been successfully loaded and authenticated, control is turned over to μRTZVisor, which resides in OCM. Upon initialization, the μRTZVisor will load and lock its complete code to the L2 cache as well as part of its critical data structures. The remaining OCM is left to be used as a shared memory area by partitions as detailed in Section 3.4. The μRTZVisor is then responsible for configuring specific hardware for the hypervisor, and for loading the guest OS images to the corresponding memory segment. Guest OSes images are not individually encrypted. As aforementioned, they are part of the overall system image. μRTZVisor does not check the integrity of the non-secure guest OSes binaries when they are loaded. This means that the chain of trust ends when the critical software is securely running. It is assumed that everything that goes outside the perimeter of the hypervisor's kernel side can be compromised. Nevertheless, the addition of another stage of verification, at the partition level, would help to achieve a supplementary level of runtime security for the entire system lifetime. By including an attestation service as a secure task, it would be possible to check and attest partition identity and integrity at boot time, as well as other key components during runtime.

3.2. Partition Manager

The main role of the partition manager is to guarantee consistency and integrity of the partitions execution context, namely their CPU state. This module also encapsulates the list of partition control blocks (PCBs), which encapsulate the state of each partition and which partition is currently active. Other kernel modules must use the partition manager interfaces to access the currently active partition and entries of the PCB for which they are responsible.

As previously explained, two types of partitions are provided: non-secure guest partitions and secure task partitions. While, for task partitions, state is limited to user mode CPU registers, for guest partitions, the state encompasses banked registers for all execution modes, non-secure world banked system registers, and co-processor state (currently, only the system configuration co-processor,

CP15, is supported). The provided VM abstraction for guest OSes is complemented by the virtualization structures of the GIC's CPU interface and distributor as well as of a virtual timer. These are detailed further ahead in Sections 3.5 and 3.8.

The partition manager also acts as the dispatcher of the system. When the scheduler decides on a different partition to execute, it informs the partition manager which is responsible for performing the context-switch operation right before leaving the kernel. In addition to saving and restoring context related to the aforementioned processor state, it coordinates the context-switching process among the different core modules, by explicitly invoking their methods. These methods save and restore partition state that they supervise, such as a memory manager method to switch between address spaces.

The partition manager also implements the delivery of asynchronous notifications to task partitions, analogous to Unix-style signals. This is done by saving register state on the task's stack and manipulating the program counter and link registers to jump to a pre-agreed point in the partition's executable memory. The link register is set to a pre-defined, read-only piece of code in the task's address space that restores its register state and jumps to the preempted instruction. The IPC manager uses this mechanism to implement the event gate abstraction (Section 3.6).

3.3. Capability Manager

The Capability Manager is responsible for mediating partitions access to system resources. It implements a capability-based access control system that enables fine-grained and flexible supervising of resources. Partitions may own capabilities, which represent a single object, that directly map to an abstract kernel concept or a hardware resource. To serve its purpose, a capability is a data structure that aggregates owner identification, object reference and permissions. The permissions field identifies a set of operations that the owning partition is allowed to perform on the referenced object. In this architecture, every hypercall is an operation over a kernel object; thus, whenever invoking the kernel, a partition must always provide the corresponding capability.

Figure 4 depicts, from a high-level perspective, the overall capability-based access control system. A partition must not be able to interact with objects for which it does not possess a capability. In addition, these must neither be forgeable or adulterable, and, thus, partitions do not have direct access to their own capabilities, which are stored in kernel space. Each partition has an associated virtual capability space, i.e., an array of capabilities through which those are accessed. Whenever performing a hypercall, the partition must provide the identifier for the target capability in the capability space, which is then translated to a capability on a global and internal capability pool. This makes it conceptually impossible for a partition to directly modify their capabilities or operate on objects for which it does not possess a capability, as only the capabilities on its capability-space are indirectly accessible. In addition, for every hypercall operation, the partition must specify the operation it wants to perform along with additional operation-specific parameters. At the kernel's hypercall entry point, the capability is checked to ensure the permission for the intended operation is set. If so, the capability manager will redirect the request to the module that implements the referenced object (e.g., for an IPC port it will redirect to IPC manager), which will then identify the operation and perform it.

At system initialization, capabilities are created and distributed according to a design-time configuration. Some capabilities are fixed, meaning that they are always in the same position of the capability space of all partitions, despite having configurable permissions. These capabilities refer to objects such as the address space, which are always created for each partition at a system's initialization. For capabilities associated with additional objects defined in the system's configuration, a name must be provided such that it is unique in a given partition's capability space. During execution, partitions can use this name to fetch the associated index in their capability space.

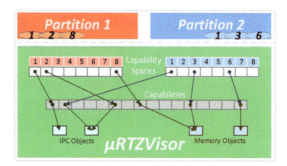

Figure 4. Capability-based access control system overview.

*μ*RTZVisor also provides mechanisms to dynamically propagate access rights by invoking Grant and Revoke operations on capabilities, which, as for any other operation, must have the respective rights set in the permissions field. The Grant operation consists of creating a derived capability, which is a copy of the original one, with only a subset of its permissions, and assigning it to another partition. The operation's recipient is notified about it, through an IPC message in one of its ports, which contains the index for the new capability in its capability space. To perform the Grant operation, the granter must specify the granting capability, the subset of permissions to grant, which must be enclosed in the original ones, and the port to which the notification will be delivered. This means that the granter must possess a capability for a port owned by the recipient. A derived capability may further be granted, giving rise to possible complex grant chains. Each derived capability is marked with a grant id, which may later be used to perform the revoke operation. In turn, the revoke operation withdraws a given capability from its owner, and can only be performed by one of the partitions in a preceding grant chain. The revocation process propagates through the donation chain. A revoked capability is maintained in a zombie state in the capability space, until it is used again. When the owning partition tries to use it, it will receive an error and the position will be freed so that it can be used again. Finally, there is a special type of capability, called a one-time capability, that can only be used once. The first time a partition uses this capability it is erased from the partition's capability space. These are also referred to as reply capabilities in the context of the IPC manager, and are leveraged to perform secure RPC communication. This is further detailed in Section 3.6.

3.4. Memory Manager

At system initialization, the memory manager starts by building the address spaces for all partitions from meta-data detailing the system image layout. For guest partitions, this encompasses figuring out which consecutive memory segments must be set as non-secure when the guest is active, i.e., those for which they were compiled to run on, and loading them to these segments. On Zynq-based devices, TrustZone memory segments have a granularity of 64 MB, which might lead to high levels of internal fragmentation. For example, for a 1 MB or 65 MB guest OS binary, 63 MB of memory is left unused. During this process, if it is detected that two guests were built to run by sharing the same memory segment, the manager will halt the system, since the spatial isolation requirement cannot be guaranteed. From the remaining free memory, the memory manager will build the virtual address spaces for secure tasks, which currently only use 1 MB pages instead of the available 4 K pages. In doing so, we simplify the manager's implementation while keeping the page table footprint low, since the latter would require second level page tables. In addition, in the current implementation, no more memory can be allocated by tasks after initialization, so partition binaries must contemplate, at compile-time, memory areas to be used as stack and heap, according to the expected system needs.

The hypervisor code and data are placed in the bottom 64 MB memory segment, which is always set as secure. Address spaces always contemplate this area as kernel memory and may extend until the

1 GB limit. Since we manage tasks' page tables, their virtual address space always starts immediately after kernel memory, making the recompilation needed for guest partitions, which always have a confined, but direct view of physical memory unnecessary. Above the 1 GB limit, the address space is fixed for all partitions, contemplating a peripheral, CPU private and system register area, and, at the top, an OCM region of TZASC 4KB segments, which we call slices and use for guest shared memory as detailed below.

This module manages two page tables used by the secure interface of the MMU. The first is a 1-to-1 mapping to physical memory and is used when a guest partition is currently active. The second is used when a task partition is active and is updated each time a new task is scheduled. Since it is expected that secure service partitions do not need to occupy a large number of pages, only the individual page table entries are saved in a list structure. The extra-overhead of updating the table at each task context restore was preferred to keeping a large and variable number of page tables and only switching the page table pointer, reducing the amount of memory used by the hypervisor. Despite the transparent view of the top physical address space, this is controlled by managing a set of three extra page tables that map the 4 KB peripheral and the TZPC and TZASC partitioning infrastructure that enables control over guest access to peripherals and shared slices. This is done in cooperation with the device manager module described in Section 3.5.

Partition memory mappings are kept active while the kernel executes, in order to avoid manual address translation for the kernel to access partition space when reading information pointed to by hypercall arguments. At the same time, the memory manager offers services to other system modules that enable them to verify the validity of the hypercall arguments (i.e., if data pointed by these arguments is indeed part of the partitions address space), read and write to and from address spaces others than the currently active one and perform address translations when needed.

Upon address space creation, a capability is inserted in the partitions' capability spaces that enables them to perform operations over it. At the moment, these operations are exclusively related to the creation and mapping of objects representing some portion of physical memory and that support shared memory mechanisms. We distinguish two different types of memory objects, page and slice objects, always represented and manipulated through capabilities and shared among partitions using grant and revoke mechanisms. Although both kinds of objects may be created by guest and task partitions, only slice objects may be mapped by guests, since guest address space control is exclusively performed through TrustZone segmentation mechanisms. For example, a guest that needs a task to process its data may create a page object representing the pages containing that data using its address space capability. It then grants the page object capability to the task. The task uses this capability to map the memory region to its own address space and processes the data. When the task is done, it signals the client partition, which revokes the capability for the page object, automatically unmapping it from the task's address space. The same can be done among tasks, and among guests, although, in the latter case, only using the slice memory region. Using this sharing mechanism, a performance boost is obtained for data processing service provision among partitions.

TrustZone-aware platforms extend secure and non-secure memory isolation to both the cache and memory translation infrastructure. Nevertheless, a cache or TLB marked as non-secure may be accessed by non-secure software, despite the current state of memory segment configuration. Hence there is a need to flush all non-secure cache lines when a new guest partition becomes active, so that they cannot access each other cached data. This is also performed for non-secure TLB entries since a translation performed by a different guest might be wrongly assumed by the MMU. This operation is not needed when switching secure tasks since TLB entries are tagged. Although it is impossible for non-secure software to access secure cache entries, the contrary is possible by marking secure page table entries as non-secure, which enables the kernel and secure tasks to access non-secure cache lines when reading hypercall arguments or using shared memory. This, however, puts forth coherency and isolation issues, which demand maintenance that negatively impacts performance. First, it becomes imperative for guest partition space to be accessed only through a memory manager's services so that it can keep track of possible guest lines pulled to the cache. When giving control back to the guest, these lines must be flushed to keep coherency in case the guest is running with caches disabled. This mechanism may be bypassed if, at configuration time, the designer can guarantee that all guest arguments are cached. Moreover, if a guest shares memory with a task, further maintenance is required, in order to guarantee that a guest with no access permission for the shared region cannot find it in the cache. This depends on the interleaved schedule of task and guests as well as the current distribution of shared memory objects.

3.5. Device Manager

The job of the device manager is similar to that of the memory manager since all peripherals are memory mapped. It encompasses managing a set of page tables and configuring TZPC registers to enable peripheral access to task and guest partitions, respectively. Each peripheral compromises a 4 KB aligned memory segment, which enables mapping and unmapping of peripherals for tasks, since this is the finest-grained page size allowed by the MMU. The peripheral page tables have a transparent and fixed mapping, but, by default, bar access to user mode. When a peripheral is assigned to a task, the entry for that device is altered to allow user mode access at each context switch. For all non attributed devices, the reverse operation is performed. An analogous operation is carried out for guests, but by setting and clearing the peripheral's secure configuration bit in a TrustZone register. If a peripheral is assigned to a partition, it can be accessed directly, in a pass-through manner, without any intervention of the hypervisor.

At initialization, the device manager also distributes device capabilities for each assigned device according to system configuration. Here, when inserting the capability in a partition's capability space, the manager automatically maps the peripheral for that partition, allowing the partition to directly interact with the peripheral without ever using the capability. However, if partitions need to share devices by granting their capabilities, the recipient must invoke a map operation on the capability before accessing the device. The original owner of the capability may later revoke it. This mechanism is analogous to the shared memory mechanism implemented by the memory manager.

There may be the need for certain devices to be shared across all guest partitions, if they are part of the provided virtual machine abstraction. These virtual devices are mapped in all guest partition spaces, being no longer considered in the aforementioned sharing and capability allocation mechanisms. A kernel module may inform the device manager that a certain device is virtual at system initialization. From there on, that module will be responsible for maintaining the state of the device by installing a callback that will be invoked at each context-switch. This is the case for the timer provided in the virtual machine, TTC1 (Triple Timer Counter 1) in the Zynq, destined to be used as a tick timer for guest OSes. We added a kernel module that maintains the illusion that the timer belongs only to the guest running on the VM by effectively freezing time when the guest is inactive. The same is true for interrupts associated with these virtual devices, and so the interrupt manager provides a similar mechanism for classifying interrupts as virtual instead of assigning them to a specific guest.

Finally, it is worth mentioning that devices themselves can be configured as secure or non-secure, which will define if they are allowed to access secure memory regions when using Direct Memory Access (DMA) facilities. We have not yet studied the intricacies of the possible interactions between secure and non-secure partitions in such scenarios, so we limit devices to make non-secure memory accesses.

3.6. IPC Manager

Communication is central to a microkernel architecture. In our approach, it is based on the notion of ports, which are kernel objects that act as endpoints through which information is read from and sent to in the form of messages. Communication mechanisms are built around the capability-system, in order to promote a secure design and enforce the principle of least authority. Thus, as explained in Section 3.3, in order to perform an IPC operation over a port, a partition must own a capability referencing that same port, with the permissions for the needed operations set. For example, a given task may own a capability for the serial port peripheral. Whenever other partitions want to interact with the device, they would need to send a message to the owning task that would interpret that message as a request, act and reply accordingly. In this scenario, a port must exist for each partition. For each port, they should possess capabilities with the minimum read and write permissions set that ensure a correct communication progression.

Port operations may work in a synchronous or asynchronous style, and are further classified as blocking or non-blocking. Synchronous communication requires that at least one partition blocks waiting for another partition to perform the complementary operation, while, in asynchronous communication, both partitions perform non-blocking operations. Synchronous communication does not require message buffering inside the kernel, improving performance since it is achieved by copying data only once, directly between partition address spaces. On the other hand, asynchronous communication requires a double data copy: first from the sender's address space to the kernel, and then from the kernel to the recipient's address space. Although this provokes performance degradation, it enforces the system's security by avoiding the asymmetric trust problem [43,44], where an untrustworthy partition may cause a server to be blocked indefinitely, preventing it from answering other partitions' requests, resulting in possible DOS attacks. This could be solved by the use of timeouts; however, there is no theory to determine reasonable timeout values in non-trivial systems [28]. In asynchronous communication, since no partition blocks waiting for another one, there is no risk of that happening. These trade-offs between performance and security must be taken into account when designing this kind of system. Despite our focus on security, we also want to offer the flexibility provided by synchronous communication, which enables fast RPC semantics for efficient service provision. As such, the μRTZVisor provides port operations for both scenarios, combining synchronous operations with the scheduling infrastructure, explained in Section 3.7.

The most elemental port operations are Send and Receive, where the former is never blocking, but the latter may be. Other operations are composed as a sequence of these elemental operations, in addition to other services provided by the scheduler, for time-slice donation, or by the capability manager for one-time capability propagation.

Table 1 summarizes all IPC primitives over available ports. As shown in the table, there are two kinds of receive operations—one blocking and the other non-blocking. In the first case, the partition will stall and wait for a complementary send operation to happen on the respective port to resume its execution. The second one will check for messages in the port's message buffer, and return one in first-in, first-out (FIFO) arrival order if available, or an error value if the port is empty.

Table 1. Port operations characterization, i.e., if it is synchronous or asynchronous and either blocking or non-blocking.

Port Operations	Synchronous	Asynchronous	Blocking	Non-Blocking
Send	x	x	–	x
RecvUnblock	-	x	–	x
RecvBlock	x	–	x	–
SendReceive	x/-	x/x	–	x
SendReceiveDonate	x/x	x/-	x	–
ReceiveDonate	x	–	x	–

If the Send operation follows a Receive that is blocking, it will happen in a synchronous style, i.e., that copy will be sent directly to the recipient's address space. Otherwise, the communication will be asynchronous, which means that a message will be pushed into the port's message buffer.

Both SendReceive operations will perform an elemental send followed by an elemental receive. In addition, they rely on services from the capability manager to grant a capability to the recipient partition. The capability may be granted permanently, like in a grant operation performed by the capability manager, or may be erased after being used once, dubbed reply-capabilities. When performing an operation with the -Donate suffix, the partition is donating its execution time-slice to the recipient partition, and it blocks its execution until receiving a response message from that same partition. More details about the donation process will be given in Section 3.7.

When a given partition asynchronously sends a message, the recipient partition may receive an event to notify it about the message arrival. Events are asynchronous notifications that alter the partition's execution flow. For the secure tasks' partitions, they are analogous to Unix signals and are implemented by the partition manager described in Section 3.2. For guest partitions, they resemble a normal interrupt, to not break the illusion provided by the virtual machine. In addition, it would be extremely difficult to implement them as signals, given that the hypervisor is OS-agnostic and has no detailed knowledge about the internals of the guest. Hence, services from the interrupt manager (Section 3.8) are used to inject a virtual interrupt in the virtual machine. To receive events, guests must configure and enable the specified interrupt in their virtual Generic Interrupt Controller (GIC). In this way, events are delivered in a model closer to the VM abstraction, and OS agnosticism is maintained. Partitions interact with the event infrastructure through a kernel object called event gate. To receive events, partitions must configure the event gate and associate it with ports that will trigger an event upon message arrival. To lower implementation complexity, each partition is assigned a single event gate, which will handle events for all ports in a queued fashion. In addition, a capability with static permissions is assigned to each partition for its event gate at system's initialization. The aforementioned permissions encompass only the `Configure` and `Finish` operations. The configure operation allows partitions to enable events, and also to specify the memory address of a data structure where event-related data will be written to upon event delivery, and that should be read by the owning partition to contextualize the event.

For synchronization purposes, there are also specific kernel objects called locks, whose operations may be blocking or unblocking. A partition may try to acquire a synchronization object by performing one of two versions of a lock: `LockBlocking` and `LockUnblocking`. The first changes partition state to blocked in case the object has already been acquired by other partition, scheduling the background domain. The latter will return the execution to the invoker, thus working like a spin-lock. To release the object, there is an `Unlock` operation. The existence of this kind of object will allow, for example, partitions to safely share a memory area for communication purposes. To do this, all of them must possess capabilities referencing the same lock object.

All communication objects must be specified at design time, which means that partitions are not allowed to dynamically create ports. In addition, the respective capabilities should be carefully distributed and configured, since, depending on its permissions, it may be possible for a partition to dynamically create new communication relations through the capability `Grant` operation, or by `SendReceive` operations. Therefore, all possible existing communication channels are, at least implicitly, defined at configuration time. Although partitions may grant port capabilities, if no relation for communication exists between partitions, they will never be able to spread permissions. Thus, designers must take care to not unknowingly create implicit channels, since isolation is reinforced by the impossibility of partitions to communicate with each other, when they are not intended to.

The port abstraction hides the identity of partitions in the communication process. They only read and write messages to and from the endpoints but do not know about the source or the recipient of those messages. This approach is safer since it hides information about the overall system structure that might be explored by malicious partitions. However, ports can be configured as privileged and messages read from these ports will contain the ID of the source partition. This enables servers to implement connection-oriented services, which might encompass several interactions, in order to distinguish amongst their clients and also to associate their internal objects with each one.

3.7. Scheduler

The presented approach for the μRTZVisor scheduler merges the ideas of [42,45]. It provides a scheduling mechanism that enables fast interaction between partitions, while enabling the coexistence of real-time and non-real-time applications without jeopardizing temporal isolation, by providing strong bandwidth guarantees.

The scheduler architecture is based on the notion of a time domain, which is an execution window with a constant and guaranteed bandwidth. Time domains are scheduled in a round-robin fashion. At design time, each time domain is assigned an execution budget and a single partition. The sum of all execution budgets constitutes an execution cycle. A partition executes in a time domain, consuming its budget until it is completely depleted, and the next time domain is then scheduled. Whenever a complete execution cycle ends, all time budgets are restored to their initial value. This guarantees that all partitions run for a specified amount of time in every execution cycle, providing a safe execution environment for time-driven real-time partitions.

The scheduler allows that multiple partitions may be assigned to a special-purpose time domain, called domain-0. Inside domain-0, partitions are scheduled in a priority-based, time-sliced manner. Furthermore, domain-0's partitions may preempt those running in different domains. It is necessary to mention that any partition is assigned a priority, which only has significance within the context of domain-0. At every scheduling point, the priorities of the currently active time domain's partition and the domain-0's highest priority ready partition are compared. If the latter possesses a higher priority, it preempts the former, but consuming its own domain's (i.e., domain-0's) time budget while executing. The preemption does not happen, of course, if domain-0 itself is the currently active domain. Figure 5 presents an example scenario containing two domains: time domain 1, assigned with partition X; and time domain 2, assigned with Partition Y, in addition to domain-0, assigned with partitions Z and W. 0 Time domain 1 is first in line, and since partition X has higher priority than the ones in domain-0, it will start to execute. After its time budget expires, a scheduling point is triggered. Time-domain 2 is next, but since domain-0's partition Z possesses higher priority than partition Y from time domain 2, Z is scheduled consuming domain-0's budget. At a certain point, partition Z blocks, and since no active partition in domain-0 has more priority than domain 2's Y, the latter is finally scheduled and executes for its time domain's budget. The next scheduling point makes domain-0 the active domain, and the only active partition, W, executes, depleting domain-0's budget. When this expires, a new execution cycle begins and domain 1's partition X is rescheduled.

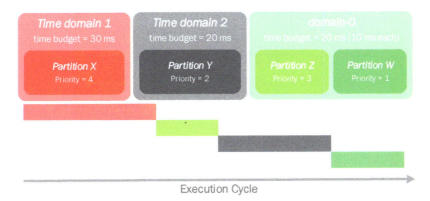

Figure 5. Example of an execution cycle, given a set of time domains with their own partitions and respective configuration.

Aiming at providing fast interaction between partitions, some IPC operations are tightly coupled with specific scheduling functionalities. Section 3.6 highlights a number of IPC operations that rely on the scheduler: the `ReceiveBlocking`, and the ones with the -donate suffix (i.e., `SendReceiveDonate` and `ReceiveDonate`). The `ReceiveBlocking` operation results in changing the partitions state to blocked, and then scheduling the next ready partition from domain-0 to perform some background work. Nevertheless, it keeps consuming the former domain's time budget, since it prevails as the active time domain. Hence, by blocking, a partition implicitly donates its execution time to domain-0. The following scheduling point will be triggered according to one of three scenarios: (a) domain-0's internal time slice expires, which results in scheduling the next highest priority partition from domain-0; (b) the active time domain's budget expires, and the next time domain becomes active; (c) the executing partition sends a message to the active time domain's partition, which would change its state to ready and result in scheduling it right away. In summary, upon blocking, a partition remains in this state until it is unblocked by receiving a message on the port it is hanging. If an execution cycle is completed without a change in the partition's state, partitions from domain-0 are scheduled once more in its place.

The -donate suffixed operations require a more intricate operation from the scheduler, and by invoking them, a partition is explicitly donating its time budget to the recipient port's owner. Hence, it will block until it has the created dependency resolved, i.e., it blocks waiting for the message's recipient to send its response. In case the donator has a higher priority than the donatee server, the latter will inherit the former's priority, augmenting the chances of it to execute and to resolve the dependency sooner. Considering a scenario where two partitions donated their time to low-priority servers running in domain-0, the server that inherits the higher priority will execute first when domain-0's becomes active, or even preempt another time domain's partition, which it previously wouldn't preempt. This enables services to be provided in a priority-based manner, i.e., maintaining the priority semantics of the requesting partitions. This priority inheritance mechanism also mitigates possible priority inversion scenarios. A partition relying on another one, and donating its time domain without any other intervener, constitutes the simplest form of a donation chain. However, a donate operation may be performed to or from a partition that is already part of a donation chain in a transitive manner, constituting a more intricate scenario. Whatever partition is at the tail of the chain, it will be the one to execute whenever one of the preceding partitions is picked by the scheduler. Nonetheless, only the one following a given partition at the donation chain is able to restore its state to ready, by sending a message to the port on which the donator is waiting for the response to its request.

This synchronous, donation-based mechanism is prone to deadlock scenarios, which in our approach is synonymous with a cycle in the donation chain. Instead of traversing the chain and looking for cycles for every donation, this problem is mitigated recurring to a bitmap for a more lightweight solution. Given that the number of partitions is fixed, each bit in the bitmap represents a partition in the system, and that, if set, means that the represented partition is in the donation chain between the respective node and the chain's tail. Every partition has at least its own bit set in its bitmap. Thus, cycles are detected whenever a match occurs by crossing both bitmaps, from the donator and the recipient for the donation.

Although not imposed by the implementation, this design was devised so that guest partitions are placed in common time-domains and secure task partitions are placed in domain-0. Since the idea of secure tasks is to encapsulate extended kernel services or shared drivers, these can be configured with lower priorities, executing according to guest needs and based on the latter's priority semantics. In addition, this models allows for the coexistence of event-driven and background partitions in domain-0, while supporting guests with real-time needs and that require a guaranteed execution bandwidth. For example, a driver with the need for a speedy reaction to interrupts could be split in two cooperating tasks: a high priority task acting as the interrupt handler, which upon interrupt-triggering would message the second lowest priority task that interfaces other partitions, executing only upon a guest request. Even though a mid-priority client guest could be interrupted by the interrupt handler, its execution time within the cycle is guaranteed. Due to possible starvation, only the tasks that act as pure servers should be configured with the lowest possible priorities in domain-0. Other partitions that may be acting as applications on their own right or may have the need to perform continuous background work should be configured with a middle range priority. It is worth mentioning that the correctness of a real-time schedule will depend on time domain budgets, partition priorities and on how partitions use communication primitives. Thus, while the hypervisor tries to provide flexible and efficient mechanisms for different real-time constraints to be met, their effectiveness will depend on the design and configuration of the system.

3.8. Interrupt Manager

Before delving into the inner workings of the interrupt manager, a more detailed explanation of TrustZone interrupt support is unavoidable. The TrustZone-aware GIC enables an interrupt to be configured as secure or non-secure. Non-secure software may access all GIC registers, but when writing bits related to secure interrupts, the operation silently fails. On the other hand, when reading a register, the bits related to secure interrupts always read as zero. Priorities are also tightly controlled. In the GIC, the priority scale is inversed, that is, low number priorities reflect the highest priorities. When secure software writes to priority registers, the hardware automatically sets the most significant bit, so non-secure interrupts are always on the least priority half of the spectrum. In addition, many of the GIC registers are banked between the two worlds. Finally, the CPU and GIC can be configured so that all secure interrupts are received in monitor mode, i.e., by the hypervisor as FIQ exceptions, and all non-secure interrupts to be directly forwarded to the non-secure world as IRQs. All of these features enable the hypervisor to have complete control over interrupts, their priority and preemption, while enabling pass-through access of guests to the GIC.

The interrupt manager works on the assumption that only one handler per interrupt exists in the system. These may reside in the kernel itself or in one of the partitions. In the first and simplest case, kernel modules register the handler with the manager at initialization time, which will be called when the interrupt occurs. At the moment, the only interrupt used by the kernel is the private timer interrupt that is used by the scheduler to time-slice time domains. If the interrupt is assigned to one partition in the configuration, the capability for the interrupt will be added to its capability space with the grant permission cleared. Partition interrupts are always initially configured as disabled and with the lowest priority possible. The details on how a interrupt is configured and handled depends on the kind of partition it is assigned to, as detailed below.

Task partitions cannot be granted access to the GIC, since, if so, by running on the secure world, they would have complete control over all interrupts. All interactions with the GIC are thus mediated by the hypervisor by invoking capability operations, such as enable or disable. These partitions receive interrupts as an IPC message. Hence, before enabling them, they must inform the interrupt manager on which port they want to receive it. The kernel will always keep task interrupts as secure, and when the interrupt takes place, it will place the message in the task's port and disable it until the task signals its completion through a hypercall. Since this process relies on IPC mechanisms, the moment in which the task takes notice of the interrupt will completely depend on what and when it uses the IPC primitives and on its scheduling priority and state. It is allowed for a task to set a priority for an interrupt, but this is truncated by a limit established during system configuration and the partitions' scheduling priority.

Guest partitions can directly interact with the GIC. For them, a virtual GIC is maintained in the virtual machine. While a guest is inactive, its interrupts are kept secure but disabled. Before running the guest, the hypervisor will restore the guest's last interrupt configurations as well as a number of other GIC registers that are banked between worlds and may be fully controlled by the guest. Active guests receive their interrupts transparently when they become pending. Otherwise, as soon as the guest becomes active, the interrupts that became pending during its inactive state are automatically triggered and are received normally through the hardware exception facilities in the non-secure world. As such, at first sight, a guest has no need to interact with the GIC or the interrupt manager through capabilities as task partitions do. However, if the capability has the right permission set, the guest can use it to signal the kernel that this interrupt is critical. If so, the interrupt is kept active while the guest is inactive, albeit with a modified priority, according to the partition's scheduling priority and a predefined configuration. Regardless of which partition is running, the kernel will receive this interrupt and manipulate the virtual GIC to set the interrupt pending as would normally happen. In addition, it will request for the scheduler to temporarily migrate the guest partition to time domain-0 (if not there already), so that it can be immediately considered for scheduling before its time domain becomes active again and handle the interrupt as fast as possible. Although the worst case interrupt latency persists as the execution cycle length minus the length of the partition's time domain, setting it as critical increases the chance of it being handled earlier, depending on the partition's priority.

Finally, in the same manner as devices, interrupts may be classified as virtual and shared among all guest partitions. These virtual interrupts are considered fixed in the virtual GIC, i.e., the currently active guest is considered the owner of the interrupt. If no guest is active, they are disabled. They are always marked as non-secure and a more intricate management of their state is needed, so as to maintain coherence to the expected behavior of the hardware. For the moment, the only virtual interrupt is the one associated with the guest's tick timer in the Zynq, TTC1.

4. Evaluation

μRTZVisor was evaluated on a Xilinx Zynq-7010 (Xilinx, San Jose, CA, USA) with a dual Arm Cortex-A9 running at 650 MHz. In spite of using a multicore hardware architecture, the current implementation only supports a single-core configuration. Our evaluation focuses on the TCB size and memory footprint imposed by the hypervisor, and the impact on guest performance related to context-switch overhead and communication throughput and latency as well as interrupt latency for both guest and task partitions. On all performed measurements, hypervisor, guest and task partitions were running with caches enabled. Both the hypervisor and partition code was compiled using the Xilinx Arm GNU toolchain (Sourcery CodeBench Lite, 16 May 2015) with -O3 optimizations.

4.1. Hypervisor and TCB Size

μRTZVisor's implementation encompassed the development of all needed drivers and libraries from scratch, so no third-party code is used. In this prototype, ignoring blank lines and comments, it compromises 6.5 K SLOC (source lines of code) from which 5.7 K are C++ and 800 are Arm assembly. This gives rise to a total 58 KB of *.text* in the final executable binary. This small TCB size, coupled with a small number of existent hypercalls (25, at the moment, although we expect this number to increase in the future), with a common entry point for capability access-control, results in a small attack surface to the hypervisor kernel. Nevertheless, we stress the fact that, for a given guest, its actual TCB size might be increased if it shows strong dependencies on non-secure task servers, which might further interact with other servers or guests. Hence, we consider that the effective TCB of a guest fluctuates according to its communication network. Furthermore, this small code size enables us to load and lock it in one of the ways of L2 cache at system initialization, resulting in increased performance and enhanced security, given this is an on-chip memory and cannot be tampered with from the outside.

Since partitions and communication channels are static, the number of structures used for their in-kernel representation and manipulation, including capabilities, are fixed at compile-time and, thus, the amount of memory used for data by the hypervisor for partition bookkeeping essentially depends on system configuration. For example, a scenario with seven partitions, two guests and five servers, where the guests share three of the servers, totaling the existence of 12 communication ports, the amount of fixed data is about 22 KB. The large majority of this data is related to capability representation, where the sum of the data used in all capability spaces is 15 KB. Capabilities are heavyweight structures mainly due to the needed tracking for grant and revoke mechanisms. We plan to refactor their implementation in the near future to reduce their memory footprint. For small setups such as the one presented above, we also load and lock this data in L2 cache; otherwise, it is kept in OCM as defined by the secure boot process (Section 3.1). As explained in Sections 3.4 and 3.5, translation table number and size are fixed. Nevertheless, these are also kept in OCM, given that the MMU does not access caches during page walks, and table updates would force a flush to the main memory. Other data structures with more volatile and dynamic characteristics such as those used for representing messages and memory objects are allocated using object pools for efficient allocation and eliminating fragmentation. These are kept in secure DRAM.

4.2. Context-Switch Overhead

To evaluate the overhead of the context-switch operation on guest virtualization, we started by running the Thread-Metric Benchmark Suite on an unmodified FreeRTOS, a widely used RTOS (Real-Time Operating System), guest. The Thread-Metric Benchmark Suite consists of a set of benchmarks specific to evaluate real-time operating systems (RTOSes) performance. The suite is comprised of seven benchmarks. For each benchmark, the score represents the RTOS impact on the running application, where higher scores correspond to a smaller impact. Benchmarks were executed in the native environment, and compared against the results when running on top of the hypervisor. The μRTZVisor was configured with a single time partition for which the period was varied between 1 and 20 ms, which effectively defines a tick for the hypervisor in this scenario. FreeRTOS was configured with a 1 millisecond tick and was assigned a dedicated timer so that time would not freeze during the context-switch as in the case of the virtual timer, allowing the guest to keep track of wall-clock time. Since only one guest was running, the hypervisor dispatcher was slightly modified to force a complete context-switch, so that results can translate the full overhead of the guest-switching operation. This includes the flushing of caches and TLB invalidation. Figure 6 presents the achieved results, corresponding to the normalized values of an average of 1000 collected samples for each benchmark. Each sample reflects the benchmark score for a 30 s execution time, encompassing a total execution time of 500 min for each benchmark. The results show that, for a 50 Hz switching rate (20 ms period), FreeRTOS performance degradation is less than 1% across all benchmarks. This degradation aggravates as the switching period decreases to 10, 5 and 1 millisecond, at which point the hypervisor's tick meets the guest's, averaging 1.8%, 3.6% and slightly less than 18%, respectively. Although not perceptible in the figure, the standard deviation of the measured benchmarks is null for the native version of FreeRTOS, while, for the virtualized ones, it increases as the time window period decreases. This appears to happen due to the fact that, if the FreeRTOS timer interrupt is triggered during the context-switch operation or the hypervisor's during the FreeRTOS tick handler, the interrupt latency or interrupt handling time drift. With a higher rate of switching operations, the occurrence of this interference increases.

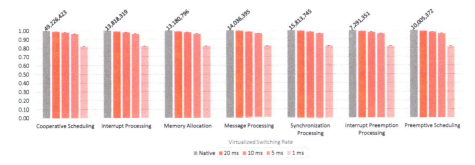

Figure 6. Thread-Metric Benchmark results, comparing the native execution of a FreeRTOS guest against its virtualized execution with different switching-rates.

Although the Thread-Metric benchmarks are useful to compare the relative performance of RTOSes by measuring common RTOS functionality performance, these are synthetic benchmarks and do not reflect the system's operation under realistic workload scenarios. We complement this assessment with the MiBench Embedded Benchmark Suite. Its 35 benchmarks are split into six suites, each targeting a specific area of the embedded market. The six categories are automotive (and industrial control), consumer devices, office automation, networking, security, and telecommunications. The benchmarks are available as standard C open-source code. We've ported the benchmarks to FreeRTOS enhanced with the FatFs file system. We focus our evaluation on the automotive suite since this is intended to assess the performance of embedded processors in embedded control systems, which we considered to be one of the target applicational areas of our system. These processors require performance in basic math abilities, bit manipulation, data input/output and simple data organization. An example of applications include air bag controllers, engine performance monitors and sensor systems. The benchmarks that emulate this scenarios include a basic math test, a bit counting test, a sorting algorithm and a shape recognition program (susan). Figure 7 shows the measured relative performance degradation for the six benchmarks in the automotive suite, each executed under a large and small input data set. The normalized column is labeled with the result for the natively executed benchmark. The benchmarks were assessed under the same conditions and hypervisor configuration as the Thread-Metric benchmarks. The results show a performance degradation of the same order of magnitude but are, nevertheless, larger than those obtained for the Thread-Metric benchmarks. A degradation of about 21.2%, 4.2%, 2% and 0.9% for the 1, 5, 10, 20 ms guest-switching rate, respectively, was obtained. We conclude that this slight increase in degradation under more realistic workloads pertains to the fact that these applications have a much larger working set than those present in the Thread-Metric Suite. Hence, the impact of a complete cache flush upon each context-switch results in a much larger cost in performance, since a much higher number of cache lines must be fetched from main memory when the guest is resumed.

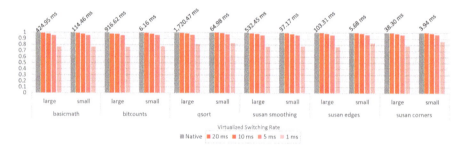

Figure 7. MiBench Automotive Benchmark results, comparing the native execution of a FreeRTOS guest against its virtualized execution with different switching-rates.

The benchmarks discussed above explain the context-switch overhead for a scenario running only guest partitions. Since guest and task partitions are represented in different kernel structures, the needed save and restore operations performed during a context-switch depend on whether the preempted partition or the incoming one are guests or tasks. All of the possible combinations give rise to four scenarios: guest–guest, task–task, guest–task and task–guest. The last scenario subdivides into two: when switching between a guest, to a task, back to the same guest, as in the case of RPC communication, there is no need for cache and TLB maintenance. However, when a guest is scheduled after a task, and the last active guest was not the same, cache flushing and TLB invalidation need to be performed. We measured the time of the context-switching operation for each scenario, running partitions in separate 10 ms time domains. Table 2 shows the average of 10,000,000 samples of each measured operation. First, we note that operations involving the invalidation of caches are an order of magnitude higher than the rest, and more than 15 times higher than the lowest one. Switching between guest and tasks or between tasks takes much less time, about 19.4 and 10.4 μs, respectively. These latter are the times involved in a brief RPC operation from a guest to a task, using the time donation facilities. This validates our initial premise that executing services in secure tasks brings great performance benefits, which is also supported by the synchronous communication results shown in Section 4.4.

Table 2. Context-switch operation execution time (µs).

Guest–Guest	166.68
Task–Task	10.38
Guest–Task	19.13
Task–Guest	19.63
Task–Different Guest	156.00

4.3. Interrupt Latency

To assess partition interrupt latency, which is defined as the time from the moment an interrupt is triggered until the final handler starts to execute, we set up a dedicated timer to trigger an interrupt every 10 ms. We measured the latency both for when the interrupt is assigned to a guest and to a task, collecting 1,000,000 samples for each case. In both cases, the scheduling configuration scenario contains two time domains in addition to domain-0, each configured with a 10 ms time window. Each time domain is assigned a different guest, and domain-0 is assigned two task partitions. The handler partition is always configured with the highest priority in the system, to translate high priority interrupt semantics when the handler partition is inactive. In addition, a domain-0 time window is enough and never completely depleted by the interrupt handler's execution, allowing the partition to react as quick as possible to interrupt events, by executing as a high priority partition in domain-0.

In the case of a guest OS partition, the interrupt is configured as critical, which means that, when the guest is inactive, the hypervisor will receive the interrupt and migrate the guest to domain-0 so that it can be immediately scheduled. Figure 8a shows the relative frequency histogram for a guest handler interrupt latencies. From the collected data, three different result regions stand out, which we identify as depending on the currently active partition at the moment the interrupt is triggered: handler guest, task or a different guest. It is clear that when the interrupt is triggered while the handler guest is executing, the measured latency is minimal since no hypervisor intervention is needed. The guest receives the interrupt transparently through normal hardware exception mechanisms and the final interrupt handler executes within an average 0.22 µs and a 0.25 µs WCET (Worst-Case Execution Time). When a different partition is active at the trigger point, the interrupt latency increases significantly since this involves the hypervisor itself receiving the interrupt, migrating the guest partition to domain-0, performing the scheduling and finally the context-switch operation. As explained in the previous section, the latter differs depending on what kind of partition is preempted, which results in an average latency of 29.03 and 180.64 µs and a WCET of 29.93 and 181.36 µ and when a task and guest are preempted, respectively. This shows that, although improving the best case scenario, placing the interrupt handler on a guest partition might result in unacceptably high latencies when a different guest needs to be preempted and caches flushed upon context-switch.

When the interrupt is handled by a task partition, the hypervisor always receives the interrupt and forwards it to the task via an IPC message. In our test scenario, the handler task is configured as the highest priority partition, executing in domain-0. Since running a high priority partition in domain-0 quickly depletes its time budget, the partition cyclically blocks on the port on which it receives the interrupt message. Figure 8b shows the relative frequency histogram with two distinct regions corresponding to the cases that result in a different task or guest preemption when the interrupt is triggered and the task unblocks by receiving the interrupt message. The results show an average 20.68 or 30.32 µs for preempted task and guest partitions, respectively. This latency arises given the extra overhead imposed by the message sending, scheduling and context-switch operations. From the presented results we conclude that, while the direct handling of an interrupt by a guest will yield a best case result, placing the interrupt handler's top half in a high priority domain-0 task (which performs the critical handling operations and notifies the bottom half executing on a different partition), will guarantee lower average and more deterministic latencies.

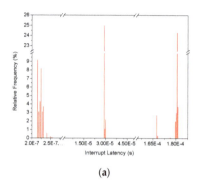
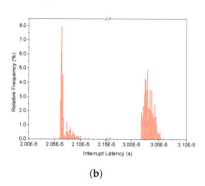

(a)   (b)

Figure 8. Relative frequency histogram of interrupt latencies (s). (**a**) guest interrupt latencies; (**b**) task interrupt latencies.

We note that, although our test scenario only contemplates tasks running in domain-0, and guests running in their own time domain, the reverse would yield similar results. Running a task in its own time domain would accrue the case on which the interrupt is triggered during the tasks execution, resulting in approximately the same latency for when a different task is preempted. A guest running in domain-0 as a high priority partition would yield the same preemption resulting latencies as those shown in Figure 8a.

4.4. IPC

To evaluate the performance of communication mechanisms, we devised scenarios for both asynchronous and synchronous communication. Table 3 shows the times needed to perform the asynchronous Send, Receive and SendReceive operations. Despite the performance varying slightly depending on whether the running partition is a guest or a secure task, this variation is not considered to be significant in case of asynchronous communication. As such, the performed measurements only reflect the time that it takes to perform the respective hypercalls from a guest partition. For a 64 byte message size, the hypercall execution time is of 4.36, 4.17 and 5.49 μs for each operation, respectively. These times increased by about 1 μs for each additional 64 bytes in the message. In all cases, there is one copy to be made. In the Send and SendReceive hypercalls, from the guest's address space into the port's message buffer, and the opposite for the Receive hypercall. There is a slight difference between the Send and SendReceive execution times, depending on the recipients' state when the operations are performed. In our implementation, this dictates whether the communication is synchronous or asynchronous. The SendReceive operation behaves similarly to the Send operation, with the addition of granting a one-time capability to the recipient guest, which only degrades performance by around 1 μs.

Table 3. Asynchronous IPC primitives latency (μs).

Message Size (bytes)	Send	Receive	Send Receive
64	4.36	4,17	5,49
128	5.17	4.75	6.31
192	6.00	5.16	7.13
256	6.82	5.72	7.98
320	7.69	6.21	8.80

In Table 4, the time measurements for synchronous IPC are presented. To infer about synchronous communication performance, we have prepared three tests, performed under three different scenarios. The first test consists on measuring an elemental Send operation, ensuring that the recipient is blocked, resulting in a synchronous operation. The second (One Way test) and the third (Two Way test) tests encompass time-slice donations procedures. In the latter, a full RPC communication cycle is measured, where the client partition waits for the server's response, while the former is used to measure only half of that cycle, i.e., the time that takes for a request to get the a blocked server. Each test scenario was performed between two guests (a); two tasks (b); and between one guest and one task (c). In each scenario, we kept the scheduling configurations simple, with only two partitions running concurrently, in this way reducing scheduling operations to the bare minimum. We note that the performance of synchronous IPC, as mentioned in Section 3.6, heavily relies on the performance of time-slice donation services provided by Scheduler and data transfer services provided by the Memory Manager, which is reflected in the achieved results. The first test evaluates the time that it takes to deliver a message to a blocked recipient, without scheduling involved. In this way, we can measure the latency introduced only by the data transfer. By comparing this value with the other two, we can measure the overhead introduced by our approach to build the donation chain and consequent scheduling operation. Whenever donating, there's a context-switch involved, and as such, the overhead imposed by this operation, as shown in Section 4.2, also applies in this context. Thus, in scenario (a); in which there are two guests communicating, results in the largest overhead in every performed test. Consequently, donation scenarios (b) and (c) involve tasks that are far more efficient. Regarding message size variation, we see that with the increase of 64 bytes in message size, the latency increases in just 1 to 2 µs, which is not too significant.

Table 4. Synchronous IPC communication latency (µs).

Message Size (bytes)	(a) Guest–Guest			(b) Task–Task			(c) Guest–Task		
	Send	One Way	Two Way	Send	One Way	Two Way	Send	One Way	Two Way
64	15.21	195.14	385.73	5.46	20.17	42.90	6.01	28.97	63.50
128	16.24	197.23	389.45	6.18	20.96	44.35	7.15	30.30	66.60
192	16.78	199.76	394.18	6.88	21.74	45.87	8.37	30.89	68.63
256	18.58	202.65	398.10	7.57	22.50	47.31	9.48	32.32	71.49
320	18.88	204.66	402.39	8.28	23.42	48.83	10.68	33.25	73.77

Synchronous communication encompasses an increased overhead, regarding, donation and scheduling operations and resulting context-switches. On the other hand, it reduces latency in service provision. As for asynchronous communication, the time involved to perform each elemental operation is smaller. Nevertheless, it should be taken into account that in client–server scenarios, the respective response could take more time than with synchronous communication, since partitions other than the server might be scheduled in the meantime. As such, the more partitions there are in a system, the bigger the latency for the response would be.

5. Related Work

Over the last few years, several embedded hypervisors have been developed in the embedded systems domain. This plethora of embedded virtualization solutions were naturally conceived following different design principles, adopting different system architectures (e.g., monolithic, microkernel, exokernel), relying on different hardware technologies (e.g., Arm VE, Arm TrustZone, Intel VT, Imagination MIPS VZ), and targeting different application domains (e.g., aerospace, automotive, industrial) [14,32–40,46–49]. Due to the extensive list of existing works under this scope, we will focus our description exclusively on TrustZone-assisted hypervisors (Section 5.1) and microkernel-based virtualization solutions (Section 5.2).

5.1. TrustZone-Assisted Virtualization

The idea of using TrustZone technology to assist virtualization in embedded systems is not new, and the first works exploiting the intrinsic virtualization capabilities of TrustZone were proposed some years ago. The majority of existing solutions just implement dual-OS support, due to the perfect match between the number of guests and the number of virtual states supported by the processors.

Winter pioneered research around the use of TrustZone technology for virtualization. In [38], Winter introduced a virtualization framework for handling non-secure world guests, and presented a prototype based on a secure version of the Linux-kernel that was able to boot only an adapted Linux kernel as non-secure world guest. Later, Cereia et al. [39] described an asymmetric virtualization layer implemented on top of the TrustZone technology in order to support the concurrent execution of both an RTOS and a GPOS (General-Purpose Operating System) on the same processor. The system was deployed and evaluated on an emulator, and never reached the light of a real hardware platform. In [35], Frenzel et al. proposes the use of TrustZone technology to implement the Nizza secure architecture. The system consists of a minimal adapted version of Linux-kernel (as normal world OS) on top of a hypervisor running on the secure world side. SafeG [40], from the TOPPERS Project [50], is a dual-OS open-source solution that takes advantage of Arm TrustZone extensions to concurrently execute an RTOS and a GPOS on the same hardware platform. Secure Automotive Software Platform (SASP) [51] is a lightweight virtualization platform based on TrustZone that consolidates a control system with a in-vehicle infotainment (IVI) system, while simultaneously guaranteeing secure device access for the consolidated automotive software. LTZVisor [36], from the TZVisor Project [52] is an open-source lightweight TrustZone-assisted hypervisor mainly targeting the consolidation of mixed-criticality systems. The hypervisor supports the coexistence of an RTOS side by side with a GPOS.

Recently, in [37], Pinto et al. presented RTZVisor. The distinct aspect of RTZVisor is the implementation of the multi-OS support. To the best of our knowledge, RTZVisor has proven to be the unique TrustZone-assisted hypervisor allowing the coexistence of multiple OSes on the same hardware platform (without VE support). Finally, VOSYSmonitor [46], developed by Virtual Open Systems, enables concurrent execution of two operating systems, such as a safety critical RTOS and a GPOS. VOSYSmonitor distinguishes from other existing works because it is implemented for Armv8-A processors, which has the option to use Arm VE for running an hypervisor such as KVM [32] , a Linux Kernel-based Virtual Machine, on the non-secure world side.

5.2. Microkernels-Based Virtualization

Microkernels were not always a viable solution, due to the architecture's reliance on IPC, which constituted a bottleneck on system performance. The L4 microkernel, developed by Jochen Liedtke, appeared to break the stigma surrounding microkernels, proving their utility when providing efficient IPC mechanisms. L4 is the root of a family tree of microkernels that have a proven record of efficient performance and reliability, by following the core idea of kernel minimality and policy-void mechanisms [28]. In this section, we briefly survey some members of this family that served as the main source of inspiration for the ideas implemented in μRTZVisor, emphasizing those which aim to support virtualization.

Fiasco is an open-source descendant of L4 implemented in C++ aimed at real-time systems. It implements protected address spaces, synchronous IPC and a scheduler with multiple fixed-priority levels, whereby kernel executes a round-robing algorithm on threads characterized with the same priority [53,54]. The latest version Fiasco.OC also includes capabilities for access-control, which are propagated through IPC [55]. Fiasco has been applied to virtualization through a para-virtualization oriented technique named OS rehosting, which aligns the kernel interface with the CPU model that is assumed by operating system kernels [56].

The NOVA microhypervisor proposes a solution that deallocates virtualization to user space, which will inherently incur performance overhead and augmented engineering effort due to the highly para-virtualized approach, although augmenting security by significantly reducing TCB's size [31]. As such, the kernel solely provides services for spacial and temporal isolation, in addition to message passing and synchronization mechanisms. In addition, kernel operations require capabilities to access the kernel objects. Capabilities are immutable, and inaccessible in user-space, thus the static system configuration prevails. Ref. [57] presents Mini-NOVA, a simplified version of NOVA ported the Arm cortex-A9 architecture from the original x86 implementation. It aims at achieving lower overhead, smaller TCB size and higher security, thus making it more flexible and portable for embedded-systems. It is worth mentioning that it has the ability to dispatch hardware tasks to virtual machines through the dynamic partial reconfiguration technology.

PikeOS is an early spin-off of the L4 microkernel, whose purpose is to address requirements of safety-critical real-time embedded systems. It features spacial and temporal isolation, favoring minimum code size, in some cases to the detriment of flexibility [8]. Its scheduling algorithm was a huge inspiration for us as explained in Section 3.7, and it aims at providing a system that enables the coexistence of time-driven and event-driven partitions. The result is not the perfect fit for this kind of system, although by properly configuring each partition, it is possible to achieve a considerably good compromise [45].

OKL4 adopts a microkernel approach completely directed at virtualization and, thus, is dubbed a microvisor [21]. It meets the efficiency of the best hypervisor and the generality and minimality of the microkernel. It provides the abstraction of a virtual machine by providing virtual CPUs, virtual MMUs and TLB, and virtual interrupts. Nevertheless, these are manipulated by guests in a para-virtualized manner, incurring performance costs. It features a fast and reliable IPC, which is abstracted by channels and virtual interrupts for synchronization purposes. It implements only asynchronous IPC, which maps better to the VM model, and is less susceptible to DOS attacks. By the heritage of its seL4 predecessor, it provides access-control facilities based on capabilities, since any kernel operation requires one. OKL4 has been augmented to take advantage of the Arm virtualization extensions and support unmodified guest OSes [33].

6. Conclusions

Modern day embedded systems are becoming increasingly complex and network-oriented. At the same time, they are supporting a number of safety and security-critical infrastructures. As such, the need for highly reliable and dependable systems consequently arises. Microkernel-based virtualization has proven to be a valid solution to guarantee functionality encapsulation and fault-containment, providing an adequate environment for mixed-critically systems, while relying on a minimal TCB. However, these systems often follow a para-virtualized approach, requiring a high engineering effort for the porting and hosting of guest OSes. Hardware-supported virtualization technologies have been widely used to mitigate these obstacles. Arm's TrustZone technology stands out given its wide presence in low to mid-range processors used in embedded systems.

This work describes μRTZVisor, a hypervisor that leverages Arm TrustZone security extensions to provide a close to fully-virtualized environment on the non-secure side of the processor. Drawing inspiration from microkernel ideas, it also allows for the execution of tasks on secure user mode, towards encapsulating extra functionality such as drivers or other services and which are accessed via IPC facilities tightly coupled with scheduling mechanisms that support fast RPC and real-time behavior. All resource accesses are wrapped by a capability-based access control mechanism, which follows a design-time configuration and statically defines partition privileges and possible communication channels. Results show that μRTZVisor presents a small TCB (approximately 60 KB) and imposes small performance overhead for isolated, unmodified guest OSes (less than 2% for a 20 ms guest-switching rate). Nevertheless, the largest hindrance of the TrustZone-based virtualization approach continues to be the need for cache flushing and TLB invalidation when supporting concurrent guests on the secure side of the processor. This leads to unacceptably high context-switching times, interrupt and communication latencies when multiple guests are involved. However, the use of secure world tasks seems to mitigate these shortcomings. Placing interrupt top halves on high priority secure tasks results in lower average and more deterministic interrupt latency. In addition, communication between guests and tasks presents much better performance especially for RPC communication taking advantage of time-slice donations. This encourages our microkernel-like approach of placing extra kernel functionality in secure tasks to guarantee a small TCB, while enabling the easy plugging of extra functionality in TrustZone-base virtualization.

Future Work

In the future, we will further study cache behavior in TrustZone-enabled platforms, to explore ideas on how to achieve better performance for multiple guest support and guest-to-guest communication. Additionally, we plan to explore available DMA facilities present on the target platforms for efficient IPC data transfer. Still regarding DMA, we also plan to study TrustZone features on compatible DMA controllers to allow for guests and tasks to use and share these facilities in a secure and isolated manner. Secure tasks are, by definition, secure kernel extensions for the provision of some shared service. The already provided isolation is desired for fault-containment, although it could incur performance degradation due to high-traffic IPC, when these services are extensively used. In the future, we would like for tasks to be configured to either be a security extension of the kernel (a secure task) or to be in the kernel itself, allowing the system's designer to decide between performance and security. In addition, we would like to maintain the IPC-like interface for both configuration scenarios, so that the service provision happens transparently to guests in the sense that, regardless of where the service is placed, accessing it happens in the same manner. The assessment of application benchmarks running on guests dependent on these services is also imperative, in order to quantify the overhead introduced by RPC in realistic workload scenarios. We would also like to port the current implementation to other Arm architectures and platforms supporting TrustZone, as well as to port other OSes to run over μRTZVisor. Analyzing the system's functioning on different contexts would allow us to identify some possible enhancements. Until now, μRTZVisor targets single-core architectures. We plan to augment our kernel to a multi-core implementation, which could result in improved real-time support and increased performance. It is also in our best interest to explore virtual machine migration, a technique widely used in server's virtualization with proven benefits [58]. To the best of our knowledge, this is a topic with little to no research in embedded systems' virtualization solutions. We think it could bring more efficient energy management through load balancing among application clusters, which we think may have applicability, for example, in industrial IoT applications. Finally, for a more thorough evaluation of our solution, we intend to compare micro and application benchmarks against other available embedded virtualization solutions (ideally open-source), under the same execution environment (i.e., same platform, same guest OS, same workload, etc.).

Acknowledgments: This work has been supported by COMPETE: POCI-01-0145-FEDER-007043 and FCT-Fundação para a Ciência e Tecnologia within the Project Scope: UID/CEC/00319/2013.

Author Contributions: Sandro Pinto conceived, designed and implemented RTZVisor. Jose Martins, Joao Alves, Adriano Tavares and Sandro Pinto conceived and designed μRTZVisor. Both Jose Martins and Joao Alves implemented μRTZVisor. Jose Martins, Joao Alves and Sandro Pinto designed and carried out the experiments. Jorge Cabral and Adriano Tavares contributed to data analysis. All authors contributed to the writing of the paper.

Conflicts of Interest: The authors declare no conflict of interest. The founding sponsors had no role in the design of the study; in the collection, analyses, or interpretation of data; in the writing of the manuscript, and in the decision to publish the results.

References

1. Heiser, G. The role of virtualization in embedded systems. In Proceedings of the ACM 1st Workshop on Isolation and Integration in Embedded Systems, Scotland, UK, 1–4 April 2008; pp. 11–16, doi:10.1145/1435458.1435461.
2. Herder, J.N.; Bos, H.; Tanenbaum, A.S. *A Lightweight Method for Building Reliable Operating Systems Despite Unreliable Device Drivers*; Technical Report IR-CS-018; Vrije Universiteit: Amsterdam, The Netherlands, 2006.
3. Moratelli, C.; Johann, S.; Neves, M.; Hessel, F. Embedded virtualization for the design of secure IoT applications. In Proceedings of the IEEE 2016 International Symposium on Rapid System Prototyping (RSP), Pittsburg, PA, USA, 6–7 October 2016; pp. 1–5, doi:10.1145/2990299.2990301.

4. Pinto, S.; Gomes, T.; Pereira, J.; Cabral, J.; Tavares, A. IIoTEED: An Enhanced, Trusted Execution Environment for Industrial IoT Edge Devices. *IEEE Internet Comput.* **2017**, *21*, 40–47, doi:10.1109/MIC.2017.17.

5. Heiser, G. Virtualizing embedded systems: Why bother? In Proceedings of the ACM 48th Design Automation Conference, San Diego, CA, USA, 5–10 June 2011; pp. 901–905, doi:10.1145/2024724.2024925.

6. Reinhardt, D.; Morgan, G. An embedded hypervisor for safety-relevant automotive E/E-systems. In Proceedings of the 2014 9th IEEE International Symposium on Industrial Embedded Systems (SIES), Pisa, Italy, 18–20 June 2014; pp. 189–198, doi:10.1109/SIES.2014.6871203.

7. Kleidermacher, D.; Kleidermacher, M. *Embedded Systems Security: Practical Methods for Safe and Secure Software and Systems Development*; Elsevier: Amsterdam, The Netherlands, 2012.

8. Kaiser, R. Complex embedded systems-A case for virtualization. In Proceedings of the IEEE 2009 Seventh Workshop on Intelligent solutions in Embedded Systems, Ancona, Italy, 25–26 June 2009; pp. 135–140.

9. Aguiar, A.; Hessel, F. Embedded systems' virtualization: The next challenge? In Proceedings of the 2010 21st IEEE International Symposium on Rapid System Prototyping (RSP), Fairfax, VA, USA, 8–11 June 2010; pp. 1–7, doi:10.1109/RSP.2010.5656430.

10. Acharya, A.; Buford, J.; Krishnaswamy, V. Phone virtualization using a microkernel hypervisor. In Proceedings of the 2009 IEEE International Conference on Internet Multimedia Services Architecture and Applications (IMSAA), Bangalore, India, 9–11 December 2009; pp. 1–6, doi:10.1109/IMSAA.2009.5439460.

11. Rudolph, L. A virtualization infrastructure that supports pervasive computing. *IEEE Perv. Comput.* **2009**, *8*, doi:10.1109/MPRV.2009.66.

12. Da Xu, L.; He, W.; Li, S. Internet of things in industries: A survey. *IEEE Trans. Ind. Inform.* **2014**, *10*, 2233–2243, doi:10.1109/TII.2014.2300753.

13. Sadeghi, A.R.; Wachsmann, C.; Waidner, M. Security and privacy challenges in industrial internet of things. In Proceedings of the 2015 52nd ACM/EDAC/IEEE Design Automation Conference (DAC), San Francisco, CA, USA, 8–12 June 2015; pp. 1–6, doi:10.1145/2744769.2747942.

14. Joe, H.; Jeong, H.; Yoon, Y.; Kim, H.; Han, S.; Jin, H.W. Full virtualizing micro hypervisor for spacecraft flight computer. In Proceedings of the 2012 IEEE/AIAA 31st Digital Avionics Systems Conference (DASC), Williamsburg, VA, USA, 14–18 October 2012; doi:10.1109/DASC.2012.6382393.

15. Shuja, J.; Gani, A.; Bilal, K.; Khan, A.U.R.; Madani, S.A.; Khan, S.U.; Zomaya, A.Y. A Survey of Mobile Device Virtualization: Taxonomy and State of the Art. *ACM Comput. Surv. (CSUR)* **2016**, *49*, 1, doi:10.1145/2897164.

16. Hohmuth, M.; Peter, M.; Hartig, H.; Shapiro, J.S. Reducing TCB size by using untrusted components: Small kernels versus virtual-machine monitors. In Proceedings of the 11th Workshop on ACM SIGOPS European Workshop ACM, Leuven, Belgium, 19–22 September 2004; p. 22, doi:10.1145/1133572.1133615.

17. Murray, D.G.; Milos, G.; Hand, S. Improving Xen security through disaggregation. In Proceedings of the ACM Fourth ACM SIGPLAN/SIGOPS International Conference on Virtual Execution Environments, Seattle, WA, USA, 5–7 March 2008; pp. 151–160, doi:10.1145/1346256.1346278.

18. Lackorzynski, A.; Warg, A.; Volp, M.; Hartig, H. Flattening hierarchical scheduling. In Proceedings of the ACM Tenth ACM International Conference on Embedded Software, New York, NY, USA, 7–12 October 2012; pp. 93–102, doi:10.1145/2380356.2380376.

19. Gu, Z.; Zhao, Q. A state-of-the-art survey on real-time issues in embedded systems virtualization. *J. Softw. Eng. Appl.* **2012**, *5*, 277, doi:10.4236/jsea.2012.54033.

20. Armand, F.; Gien, M. A practical look at micro-kernels and virtual machine monitors. In Proceedings of the 6th IEEE Consumer Communications and Networking Conference (CCNC 2009), Las Vegas, NV, USA, 10–13 January 2009; pp. 1–7, doi:10.1109/CCNC.2009.4784874.

21. Heiser, G.; Leslie, B. The OKL4 Microvisor: Convergence point of microkernels and hypervisors. In Proceedings of the ACM First ACM Asia-Pacific Workshop on Workshop on Systems, New Delhi, India, 30 August 2010; pp. 19–24, doi:10.1145/1851276.1851282.

22. Heiser, G. Secure embedded systems need microkernels. *USENIX Login* **2005**, *30*, 9–13.

23. Tanenbaum, A.S.; Herder, J.N.; Bos, H. Can we make operating systems reliable and secure? *Computer* **2006**, *39*, 44–51, doi:10.1109/MC.2006.156.

24. Kuz, I.; Liu, Y.; Gorton, I.; Heiser, G. CAmkES: A component model for secure microkernel-based embedded systems. *J. Syst. Softw.* **2007**, *80*, 687–699, doi:10.1016/j.jss.2006.08.039.

25. Herder, J.N.; Bos, H.; Gras, B.; Homburg, P.; Tanenbaum, A.S. Modular system programming in MINIX 3. *USENIX Login* **2006**, *31*, 19–28.

26. Hartig, H.; Hohmuth, M.; Liedtke, J.; Wolter, J.; Schonberg, S. The performance of μ-kernel-based systems. In Proceedings of the ACM SIGOPS Operating Systems Review, Saint Malo, France, 5–8 October 1997; Volume 31, pp. 66–77, doi:10.1145/268998.266660.

27. Leslie, B.; Van Schaik, C.; Heiser, G. Wombat: A portable user-mode Linux for embedded systems. In Proceedings of the 6th Linux.Conf.Au, Canberra, Australia, 18–23 April 2005; Volume 20.

28. Elphinstone, K.; Heiser, G. From L3 to seL4 what have we learnt in 20 years of L4 microkernels? In Proceedings of the Twenty-Fourth ACM Symposium on Operating Systems Principles ACM, New York, NY, USA, 3–6 November 2013; pp. 133–150, doi:10.1145/2517349.2522720.

29. Ruocco, S. A real-time programmer's tour of general-purpose L4 microkernels. *EURASIP J. Embed. Syst.* **2007**, *2008*, 234710, doi:10.1155/2008/234710.

30. Uhlig, R.; Neiger, G.; Rodgers, D.; Santoni, A.L.; Martins, F.C.; Anderson, A.V.; Bennett, S.M.; Kagi, A.; Leung, F.H.; Smith, L. Intel virtualization technology. *Computer* **2005**, *38*, 48–56, doi:10.1109/MC.2005.163.

31. Steinberg, U.; Kauer, B. NOVA: A microhypervisor-based secure virtualization architecture. In Proceedings of the ACM 5th European Conference on Computer Systems, New York, NY, USA, 13–16 April 2010; pp. 209–222, doi:10.1145/1755913.1755935.

32. Dall, C.; Nieh, J. KVM/ARM: The design and implementation of the linux ARM hypervisor. In *ACM Sigplan Notices*; ACM: New York, NY, USA, 2014; Volume 49, pp. 333–348, doi:10.1145/2541940.2541946.

33. Varanasi, P.; Heiser, G. Hardware-supported virtualization on ARM. In Proceedings of the ACM Second Asia-Pacific Workshop on Systems, Shanghai, China, 11–12 July 2011; p. 11, doi:10.1145/2103799.2103813.

34. Zampiva, S.; Moratelli, C.; Hessel, F. A hypervisor approach with real-time support to the mips m5150 processor. In Proceedings of the IEEE 2015 16th International Symposium on Quality Electronic Design (ISQED), Santa Clara, CA, USA, 2–4 March 2015; pp. 495–501, doi:10.1109/ISQED.2015.7085475.

35. Frenzel, T.; Lackorzynski, A.; Warg, A.; Härtig, H. Arm trustzone as a virtualization technique in embedded systems. In Proceedings of the Twelfth Real-Time Linux Workshop, Nairobi, Kenya, 25–27 October 2010.

36. Pinto, S.; Pereira, J.; Gomes, T.; Tavares, A.; Cabral, J. LTZVisor: TrustZone is the Key. In Proceedings of the 29th Euromicro Conference on Real-Time Systems (Leibniz International Proceedings in Informatics), Dubrovnik, Croatia, 28–30 June 2017; Bertogna, M., Ed.; Schloss Dagstuhl—Leibniz-Zentrum fuer Informatik: Dagstuhl, Germany; Volume 76, doi:10.4230/LIPIcs.ECRTS.2017.4.

37. Pinto, S.; Pereira, J.; Gomes, T.; Ekpanyapong, M.; Tavares, A. Towards a TrustZone-assisted Hypervisor for Real Time Embedded Systems. *IEEE Comput. Archit. Lett.* **2016**, doi:10.1109/LCA.2016.2617308.

38. Winter, J. Trusted computing building blocks for embedded linux-based ARM Trustzone platforms. In Proceedings of the 3rd ACM Workshop on Scalable Trusted Computing, Alexandria, VA, USA, 31 October 2008; pp. 21–30, doi:10.1145/1456455.1456460.

39. Cereia, M.; Bertolotti, I.C. Virtual machines for distributed real-time systems. *Comput. Stand. Interfaces* **2009**, *31*, 30–39, doi:10.1016/j.csi.2007.10.010.

40. Sangorrin, D.; Honda, S.; Takada, H. Dual operating system architecture for real-time embedded systems. In Proceedings of the 6th International Workshop on Operating Systems Platforms for Embedded Real-Time Applications (OSPERT), Brussels, Belgium, 7–9 July 2010; pp. 6–15.

41. Alves, T.; Felton, D. TrustZone: Integrated Hardware and Software Security. *Technol. Depth* **2004**, *3*, 18–24.

42. Steinberg, U.; Wolter, J.; Hartig, H. Fast component interaction for real-time systems. In Proceedings of the 17th Euromicro Conference on Real-Time Systems, (ECRTS 2005), Washington, DC, USA, 6–8 July 2005; pp. 89–97, doi:10.1109/ECRTS.2005.16.

43. Herder, J.N.; Bos, H.; Gras, B.; Homburg, P.; Tanenbaum, A.S. Countering ipc threats in multiserver operating systems (a fundamental requirement for dependability). In Proceedings of the 2008 14th IEEE Pacific Rim International Symposium on Dependable Computing (PRDC'08), Taipei, Taiwan, 15–17 December 2008; pp. 112–121, doi:10.1109/PRDC.2008.25.

44. Shapiro, J.S. Vulnerabilities in synchronous IPC designs. In Proceedings of the IEEE 2003 Symposium on Security and Privacy, Berkeley, CA, USA, 11–14 May 2003; pp. 251–262, doi:10.1109/SECPRI.2003.1199341.

45. Kaiser, R.; Wagner, S. Evolution of the PikeOS microkernel. In Proceedings of the First International Workshop on Microkernels for Embedded Systems, Sydney, Australia, 16 January 2007; p. 50.

46. Lucas, P.; Chappuis, K.; Paolino, M.; Dagieu, N.; Raho, D. VOSYSmonitor, a Low Latency Monitor Layer for Mixed-Criticality Systems on ARMv8-A. In Proceedings of the 29th Euromicro Conference on Real-Time Systems (Leibniz International Proceedings in Informatics), Dubrovnik, Croatia, 28–30 June 2017; Marko, B., Ed.; Schloss Dagstuhl—Leibniz-Zentrum fuer Informatik: Dagstuhl, Germany, 2017; Volume 76; doi:10.4230/LIPIcs.ECRTS.2017.6.

47. Masmano, M.; Ripoll, I.; Crespo, A.; Metge, J. Xtratum: A hypervisor for safety critical embedded systems. In Proceedings of the 11th Real-Time Linux Workshop, Dresden, Germany, 28–30 September 2009; pp. 263–272.

48. Ramsauer, R.; Kiszka, J.; Lohmann, D.; Mauerer, W. Look Mum, no VM Exits! (Almost). In Proceedings of the 13th International Workshop on Operating Systems Platforms for Embedded Real-Time Applications (OSPERT), Dubrovnik, Croatia, 14 April 2017.

49. Pak, E.; Lim, D.; Ha, Y.M.; Kim, T. Shared Resource Partitioning in an RTOS. In Proceedings of the 13th International Workshop on Operating Systems Platforms for Embedded Real-Time Applications (OSPERT), Dubrovnik, Croatia, 14 April 2017.

50. Toppers.jp. Introduction to the SafeG. 2017. Available online: http://www.toppers.jp/en/safeg.html (accessed on 29 September 2017).

51. Kim, S.W.; Lee, C.; Jeon, M.; Kwon, H.; Lee, H.W.; Yoo, C. Secure device access for automotive software. In Proceedings of the IEEE 2013 International Conference on Connected Vehicles and Expo (ICCVE), Las Vegas, NV, USA, 2–6 December 2013; pp. 177–181, doi:10.1109/ICCVE.2013.6799789.

52. Tzvisor.org. TZvisor—TrustZone-assisted Hypervisor. 2007. Available online: http://www.tzvisor.org (accessed on 29 September 2017).

53. Schierboom, E.G.H. Verification of Fiasco's IPC Implementation. Master's Thesis, Computing Science Department, Radboud University, Nijmegen, The Netherlands, 2007.

54. Steinberg, U. Quality-Assuring Scheduling in the Fiasco Microkernel. Master's Thesis, Dresden University of Technology, Dresden, Germany, 2004.

55. Smejkal, T.; Lackorzynski, A.; Engel, B.; Völp, M. Transactional IPC in Fiasco.OC. In Proceedings of the 11th International Workshop on Operating Systems Platforms for Embedded Real-Time Applications (OSPERT), Lund, Sweden, 7–10 July 2015; pp. 19–24.

56. Lackorzynski, A.; Warg, A.; Peter, M. Virtual processors as kernel interface. In Proceedings of the Twelfth Real-Time Linux Workshop, Nairobi, Kenya, 25–27 October 2010.

57. Xia, T.; Prévotet, J.C.; Nouvel, F. Mini-nova: A lightweight arm-based virtualization microkernel supporting dynamic partial reconfiguration. In Proceedings of the 2015 IEEE International Parallel and Distributed Processing Symposium Workshop (IPDPSW), Hyderabad, India, 25–29 May 2015; pp. 71–80, doi:10.1109/IPDPSW.2015.72.

58. Voorsluys, W.; Broberg, J.; Venugopal, S.; Buyya, R. Cost of Virtual Machine Live Migration in Clouds: A Performance Evaluation. *Cloud Com* **2009**, *9*, 254–265.

Article

An Energy Box in a Cloud-Based Architecture for Autonomous Demand Response of Prosumers and Prosumages

Giovanni Brusco, Alessandro Burgio * , Daniele Menniti, Anna Pinnarelli, Nicola Sorrentino and Luigi Scarcello

Department of Mechanical, Energy and Management Engineering, University of Calabria, 87036 Rende, Italy; giovanni.brusco@unical.it (G.B.); daniele.menniti@unical.it (D.M.); anna.pinnarelli@unical.it (A.P.); nicola.sorrentino@unical.it (N.S.); luigi.scarcello@unical.it (L.S.)
* Correspondence: alessandro.burgio@unical.it; Tel.: +39-0984-494707

Received: 11 September 2017; Accepted: 9 November 2017; Published: 16 November 2017

Abstract: The interest in the implementation of demand response programs for domestic customers within the framework of smart grids is increasing, both from the point of view of scientific research and from the point of view of real applications through pilot projects. A fundamental element of any demand response program is the introduction at customer level of a device, generally named energy box, able to allow interaction between customers and the aggregator. This paper proposes two laboratory prototypes of a low-cost energy box, suitable for cloud-based architectures for autonomous demand response of prosumers and prosumages. Details on how these two prototypes have been designed and built are provided in the paper. Both prototypes are tested in the laboratory along with a demonstration panel of a residential unit, equipped with a real home automation system. Laboratory tests demonstrate the feasibility of the proposed prototypes and their capability in executing the customers' loads scheduling returned by the solution of the demand-response problem. A personal computer and Matlab software implement the operation of the aggregator, i.e., the intermediary of the energy-integrated community constituted by the customers themselves, who participate in the demand response program.

Keywords: prosumer problem; scheduling; residential appliances; demand response; renewable sources; home automation system

---

1. Introduction

Many attractive definitions are currently used for demand response (DR); two of them are: "today's killer app for smart grid" and "the key for engaging consumers in the smart grid". A conventional definition states that DR is a program of actions, aimed at changes in habits of end-use customers in consuming energy [1]; these changes are mainly driven by price signals and discounted rates that discourage the use of electricity during peak hours and allow savings on bills [2]. The easiest way to induce a customer to change his habits in using electricity is to schedule its electric loads; the bigger the savings in the bill achieved through loads scheduling, the higher the participation of the customer in a DR program. A comprehensive view on technical methodologies and architecture, and socioeconomic and regulatory factors that could facilitate the uptake of DR is reported in [3]. In the path of change mentioned above, the customer does not act as an individual; on the contrary, the customer is a member of an integrated community, coordinated by an intermediary named aggregator or coalition coordinator [4–7].

In the framework of this integrated community, the aggregator and the customer communicate with each other very frequently; the energy box (EB) is the device, which interfaces the aggregator and the customer [3]. For instance, in [8], the EB is the instrument placing bids on the Portuguese tertiary market. The aggregator negotiates with the grid operator, placing a bid, which depends on bids previously returned by all EBs. Each EB can provide two different bids where the first pays for a change in the state of charge of batteries—optimally sized as in [9] and placed in the residential unit—whereas the second pays for a change in the temperature set point of air conditioning of the residential unit. In [10], the EB is a full energy home controller able to manage between five and ten loads; two Linux-based EBs mounted with a ZigBee controller for the Spanish and French trial are presented. Also in [11], the EB is a home controller for a cost-effective nonintrusive load monitor, which provides a mechanism to easily understand and control energy consumption as well as to securely permit reconfiguration and programming from anywhere. On the contrary, the EB presented in [12] is a sophisticated software for energy management systems, consisting of a suite of algorithms which support the sequential decision-making process under uncertainty in a stochastic dynamic programming network. Also, the EB presented in [13] is a software; more precisely, it is a unique decision-making algorithm which serves an automated DR program where the automated characteristic aims to reduce uncertainty and complexity of daily price fluctuation from the customer's perspective. Lastly, the EB proposed in [14] provides the wind power measurements via a multilayer feed-forward back-propagation neural network.

Given the aforementioned, the design of an EB is a hard task and full of challenges; a first and well-known challenge is to design the EB as a useful tool for communication between the consumer and the aggregator. A further challenge is to design an EB able to interact with the home energy management system (HEMS); this is because an HEMS, in general, does not necessarily share information and data (e.g., energy consumption measurements) with third parties, and it does not necessarily accept commands (e.g., turn on/off appliances) from third parties. In a residential unit, the HEMS is the device that monitors, controls and optimizes all peripherals, storage systems, distributed generators and smart meters, as illustrated in Figure 1. The HEMS asks the smart meters for energy consumptions so as to build a detailed and comprehensive status of the residential unit. Such a status is useful to precisely model and estimate energy consumption of the residential unit by using, as an example, a learning-based mechanism as in [15,16]. Therefore, the interoperability between EB and HEMS is a crucial point for implementation of a DR program in a framework of integrated community. Interoperability must be ensured regarding the communication protocols, the data format and the physical media used for communication; interoperability must also be extended to smart meters and all peripherals [17] which operate in islanded mode or in absence of a HEMS. When the same vendor produces the EB, the HEMS, the smart meters, the smart plugs and peripherals, the full interoperability is implicitly achieved; on the contrary, the management of devices produced by different vendors is much more difficult. A last challenge is to design the EB as a cost-effective tool if this technology is intended for a large number of customers, including less affluent people.

This paper proposes a new EB as a viable solution to the challenge of the communication between consumer and aggregator, and to the challenge of the interaction between an EB and HEMS. The paper also illustrates two prototypes of the proposed EB to have a clear overview of management and material requirements, as well as of cost-effectiveness; both prototypes are tested in the lab using a real home automation system.

Concerning the communication between the consumer and the aggregator, the proposed EB communicates with the aggregator's IT platform and requires access to an application, service or system [18]. Communication is over the Internet and is based on the traditional client–server paradigm; HyperText Transfer Protocol over Secure Socket Layer (HTTPS) is the communication protocol used for secure communication. As illustrated in Figure 1, the proposed EB connects to the aggregator, uploads the user's preferences and requires access to the application called prosumer problem; the application responds by providing the optimal loads scheduling. Furthermore, as in Figure 1, the EB uploads power/voltage/current measurements and it requires access to the service called monitoring and calculation; the service responds by providing reports and statistics.

Concerning the interaction with a HEMS, the proposed EB overcomes the problem of interaction because it directly communicates with all peripherals of the home automation system, bypassing the HEMS as in Figure 1. In particular, the proposed EB generates the control frames to turn on/off a load or to acquire a measurement from a meter; then, the EB sends these frames to loads and meters via the physical media for the communication of the home automation system, i.e., the typical shielded-twisted pair cable. Since the proposed EB bypasses the HEMS, two significant benefits are achieved. The first advantage is to overcome the obstacles placed by the HEMSs currently available on the market regarding the connection to devices supplied by third parties. Indeed, while HEMS are equipped with USB, RJ45 or RS482 ports for a peer-to-peer connection to EBs, a time-consuming configuration of both of these devices is usually necessary, in absence of any guarantee of obtaining a proper interaction and interoperability. The second advantage is to provide the DR program with the diagnostic, monitoring and control functions. While very important, these functions are not performed by the HEMSs currently available on the market or they are performed with insufficient quality. As a remedy, the proposed EB can be successfully adopted because it is unchained to HEMS and the valuable functions implemented by the EB are not stifled by the limits of the HEMS itself. In particular, the proposed EB uses easy programming languages to enable the expert customer to autonomously run a DR program and implement new functions and procedures that best meet its own needs.

Concerning the management and material requirements also in terms of cost-effectiveness, the two prototypes of the proposed EB were designed and tested in the laboratory; tests were carried out in conjunction with a demonstration panel of a residential unit, equipped with a real home automation system by Schneider Electric. The first prototype has a limited calculation capability and a low-cost (low-EB), the second prototype has a higher cost but also a higher capacity for solving calculation problems (high-EB). The low-EB prototype communicates with the aggregator over the internet to exchange data and submit service requests; in particular, it asks and receives from the aggregator the optimum scheduling of the consumer's loads. The low-EB prototype applies scheduling without the help of HEMS by sending the on/off commands to the peripherals of the automation system. The high-EB prototype performs the same functions as the low-EB prototype, and in addition, it is capable of calculating optimum scheduling of consumer loads.

This paper is organized as follows. Section 2 illustrates a home energy management system problem, namely "prosumer problem"; it optimizes the loads scheduling of a customer taking into account the distributed generators, the energy storage systems, the daily fluctuation in electricity prices and loads demand. Section 3 illustrates the case study of a grid-connected customer with a 3 kW PV plant and a 0.8 kW wind turbine; the customer is also equipped with a 3–6 kWh lithium-ion battery pack. Finally, Section 4 anticipates the conclusions and illustrates two laboratory prototypes of the proposed EB, namely low-EB and high-EB. The first of two prototypes uses an Arduino MEGA 2560, while the second prototype uses a Raspberry Pi3.

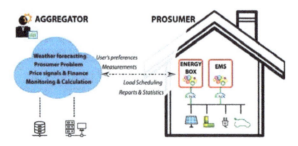

Figure 1. An energy box in a cloud-based architecture for autonomous demand response. HEMS: home energy management system.

2. The Prosumer Problem

A home energy management system problem, henceforth referred to simply as «prosumer problem» and previously presented by the authors in [19], is illustrated in this section. The prosumer problem optimizes the operation of domestic appliances taking into account the daily fluctuation in electricity prices, an electric energy storage system of electrochemical cells, the power generated by distributed generators exploiting renewable energy sources (RESs) and by other programmable generators. Nonprogrammable distributed generators exploiting RESs are a photovoltaic (PV) plant and a micro wind turbine (WT); their 24 h production forecast is assumed to be available. In addition, a biomass boiler mounted with a free-piston Stirling engine and a linear electric generator realizes a micro combined heat and power generator (mCHP); this generator provides thermal power in the form of hot water and electric power [20]. It represents a programmable generator. The electric loads are divided into two groups, group A and group B. Schedulable loads and curtailable loads belong to group A; the lower case letter a is the index over this group. Non-controllable loads belong to group B; and the lower case letter b is the index over this group. As illustrated in Figure 2, the operating cycle of a non-controllable load is not subject to changes by third parties, therefore, this cycle is an input data for the prosumer problem. On the contrary, the operating cycle of a schedulable load can be entirely shifted in the time interval $[\alpha, \beta]$, whereas the operating cycle of a curtailable load can be entirely or partially shifted in the time interval $[\alpha, \beta]$. An optimization problem encapsulates the prosumer problem to return the load scheduling which minimizes the daily electricity cost:

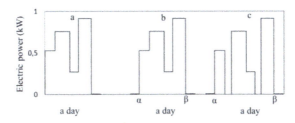

Figure 2. (**a**) Non-controllable; (**b**) schedulable; (**c**) curtailable loads.

$$min \sum_{h=1}^{24} \left\{ c_{imp}{}^h \cdot P_{imp}^h \cdot \Delta t - c_{exp}{}^h \cdot P_{exp}^h \cdot \Delta t \right\}, \tag{1}$$

subject to:

$$P_{imp}^h + P_{dis}^h + P_{WT}{}^h + P_{PV}{}^h + P_{mCHP}{}^h = P_{exp}^h + P_{cha}^h + \sum_{a=1}^{A} y_a^h \cdot Pload_a + \sum_{b=1}^{B} Pload_b^h, \tag{2}$$

$$\sum_{h=\alpha_a}^{\beta_a} y_a^h = \theta_a, \tag{3}$$

$$\sum_{h=\alpha_a}^{\beta_a - \theta_a + 1} z_a^h = 1, \tag{4}$$

$$z_a^h = 0 \ \forall h \in H - [\alpha_a, \beta_a - \theta_a + 1], \tag{5}$$

$$y_a^h \geq z_a^h \forall h \in [\alpha_a, \beta_a], \tag{6}$$

$$y_a^{h+1} \geq z_a^h \ \forall h \in [\alpha_a, \beta_a], \tag{7}$$

$$y_a^{h+\theta_a-1} \geq z_a^h \ \forall h \in [\alpha_a, \beta_a], \tag{8}$$

$$0 \leq P_{imp}^h, \ P_{exp}^h \leq P_{committed}, \tag{9}$$

$$0 \leq P_{cha}^h \leq m \cdot P_{batt}, \tag{10}$$

$$0 \leq P_{dis}^h \leq P_{batt}, \tag{11}$$

$$E_{STO} + \sum_{i=0}^{h} \left(\eta \cdot P_{cha}^i - P_{dis}^i \right) \cdot \Delta t \geq \frac{SOC_{min}}{100} \cdot C, \tag{12}$$

$$E_{STO} + \sum_{i=0}^{h} \left(\eta \cdot P_{cha}^i - P_{dis}^i \right) \cdot \Delta t \leq \frac{SOC_{max}}{100} \cdot C, \tag{13}$$

$$m = recharge\_current/discharge\_current, \tag{14}$$

$$h \in [1 \ldots 24], \tag{15}$$

Equation (1) is the cost function, which calculates the daily net electricity cost as the difference between the import price c_{imp}^h multiplied by the imported energy $P_{imp}^h \cdot \Delta t$ and the export price c_{exp}^h multiplied by the exported $P_{exp}^h \cdot \Delta t$. Equation (2) is the power balance and imposes equality between the generated powers (exported to the grid, supplied by the batteries, produced by distributed generators) and the requested ones (imported from the grid, fed into the batteries, demanded by schedulable loads, demanded by non-controllable loads). P_{PV}^h, P_{WT}^h and P_{mCHP}^h are the 24 h power generation forecast for the photovoltaic system, the micro wind turbine and the mCHP, respectively. The binary variable y_a^h is 1 when the a-th schedulable load is turned ON at the h-th hour, it is zero and vice versa. Similarly, $P_{load_b}^h$ is the load demand of the b-th non-controllable schedulable load at the h-th hour. Equation (3) sets the duration of the operating cycle of each schedulable load, whereas Equations (4) and (5) define the start time of the operating cycle. Equations (6)–(8) ensure that the operating cycle is not divided into sub-cycles for non-curtailable loads. Equation (9) constrains the power flow between the grid and the prosumer at the point of delivery; power flow is upper bounded by the committed power. Equations (10) and (11) constrain the battery charge and discharge power flow. In particular, Equation (10) limits the recharge power P_{cha}^h up to the batteries' rated power multiplied by the coefficient m, defined in Equation (14) as the ratio between the recharge current and the discharge current. Equations (12) and (13) calculate the hourly state of charge (SOC) of the batteries, taking into account the state of charge at the beginning of the day, i.e., E_{STO}. Equation (14) returns the coefficient m as the ratio between the recharge and the discharge battery current; evidently, this coefficient takes into account battery technologies and the ability to charge the batteries with the current m times the discharge one. As an example, m is higher than 1 when Li-ion batteries are used, and m is lower than 1 when lead-acid batteries are used. The input data of the prosumer problem are:

- the power forecast for generators, P_{PV}^h and P_{WT}^h;
- the hourly electricity prices, $Cimp^h$ and $Cexp^h$;
- the load profiles of non-controllable loads, $Pload_b^h$;
- the customer's parameters, α_a and β_a.

3. The Case Study

The case study is a residential unit connected to the low voltage (230 V/50 Hz) distribution grid and equipped with distributed generators, an electric energy storage system and a home automation system.

3.1. Household Appliances and Electricity Demand

We considered common household appliances: internal and external lightings, a personal computer, a TV set, a refrigerator, an air-conditioning system, a washing machine, a tumble dryer, a dishwasher and an electric vehicle. In the case that the householder does not participate in any DR program, the electricity usage is the result of familiar habits. So, taking into account the statistical surveys reported in [21–23], the considered load demand profile is that reported in Figure 3 with a dotted line. The daily energy consumption is 24.97 kWh. The peak load is 4.31 kW at 21:00; the daily average load is 1.05 kW when the peak load value is considered, it decreases to 0.81 kW and vice versa.

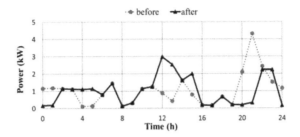

Figure 3. Load demand before and after the prosumer problem solution.

3.2. Distributed Generators

Two distributed generators which exploit renewable energy sources contribute to satisfying the load demand; the first generator is a 3 kWp photovoltaic (PV) plant on the rooftop of a residential unit, the second one is a 0.8 kW wind turbine (WT) placed in the garden. The profile of the power generated by the PV plant and WT on 1 July 2015 are in Figure 4; real data measured in southern Italy were considered. The PV plant and the WT supply the local loads, then the batteries, and finally export into the grid.

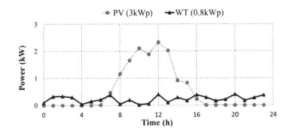

Figure 4. Twenty-four-hour photovoltaic system and wind turbine power production. PV: photovoltaic; WT: wind turbine.

3.3. Electric Energy Storage System

An electric energy storage system serves the residential unit so as to increase self-consumption. Among the storage technologies, batteries have been preferred due to their modularity and drive simplicity. In this paper lithium-ion batteries with a peak power of 3 kW and a capacity of 6 kWh are considered; the charge current is one time the discharge current. The round trip efficiency is 80%, the depth-of-discharge (SOC_{min}) is 30% and the depth-of-charge (SOC_{max}) is 98%. The self-discharge rate has been neglected. Since, as a hypothesis, the householder does not participate in a DR program, the PV plant and the WT recharge batteries when their generated power exceeds the load demand; the grid does not recharge or discharge the batteries. The state of charge of the batteries on 1 July 2015 is shown in Figure 5. The state of charge at midnight equals the SOC_{min} and batteries are temporally inoperative until the morning; from 8:00 to 12:00 the state of charge increases almost linearly from 30% up to 98%. From 12:00 to 20:00, the state of charge slightly changes and remains in the range of 75–98%; then it rapidly decreases to 30% in one hour.

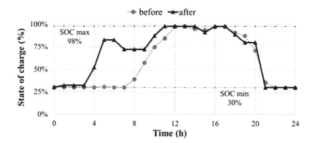

Figure 5. State of charge (*SOC*) before and after the prosumer problem solution.

3.4. Imported and Exported Electricity, Prices and Bill

Let us consider that an electricity retailer supplies power to the residential unit with a committed power equal to 6 kW. Given the load demand, the generation of distributed generators and the batteries, the power flow profile at the point of delivery is shown in Figure 6. Due to the sales contract signed between the retailer and the householder, both the import and the export prices change hourly as shown in Figure 7. The import price is calculated as the hourly wholesale market price plus a spread of 5% whereas the export price is calculated as the hourly zonal price minus a spread of 4% [24]. On 1 July 2015, the householder pays 0.52 € for energy consumption and power losses; on the other hand, he receives 0.15 € for exported energy. The daily net cost is 0.37 €.

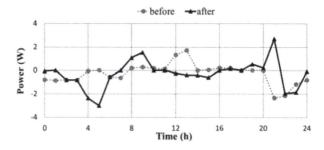

Figure 6. Power at the point of delivery before and after the prosumer problem solution.

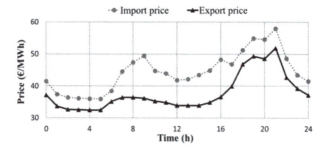

Figure 7. Hourly import and export prices.

3.5. Load Sharing between Distributed Generators and the Grid

The load demand is entirely satisfied by the distributed generators and the grid, without interruptions. With reference to Figure 8, the grid provides 40.28% of load demand while the distributed generators provide the remaining part. For the latter, 21.11% is directly provided by PV plant to loads, 20.06% is directly provided by the WT to loads, and the remaining 18.55% is indirectly provided through the batteries.

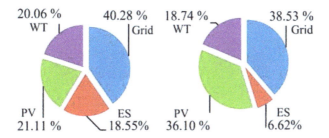

Figure 8. Load sharing between distributed generators and grid after the prosumer problem solution. ES: energy storage.

3.6. The Prosumer Problem Solution

As stated previously, for the householder who does not participate in any DR program, the electricity usage is the result of familiar habits: typically, family members leave the house in the morning and they return in the late afternoon. The electricity usage is the result of the load scheduling shown in Figure 9. Considering that the householder decides to participate in a DR program, household appliances are divided into the three load categories of Figure 2, namely non-controllable, shiftable and curtailable loads. Non-controllable loads include internal and external lightings, the personal computer, the TV set, the refrigerator and the air-conditioning system. Shiftable loads are: the washing machine (WM) and the tumble dryer (TD). Curtailable loads are: the dishwasher (DW) and the electric vehicle (EV). Being curtailable, the operating cycle of the DW and the EV can be divided in sub-cycles where each sub-cycle lasts 15 min or integer multiples of 15 min. It is worth noting that the tumble drier must start within 2 h of the end of the washing machine cycle. Shiftable loads parameters are collected in Table 1 where the last column reports the family members' preference about the time interval, from α_a to β_a available to execute the entire operating cycle of each shiftable load. A fundamental hypothesis is that the accession of the householder to the DR program allows the aggregator to remotely manage the batteries placed in the residential unit. Therefore, the aggregator may require power exchanges between the batteries and the grid in order to achieve bill savings.

Table 1. Shiftable appliances parameters.

Appliance	Power (kW)	Op. Cycle (min)	Curtailable	Time Interval α_a–β_a
Washing machine	2.1	120	no	21:00–19:00
Tumble dryer	1.2	60	no	21:00–19:00
Dish washer	1.9	120	yes	20:00–06:00
Electric vehicle	0.0 ÷ 3.0	0 ÷ 300	yes	19:00–07:00

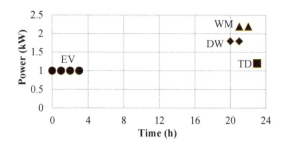

Figure 9. Load scheduling according to the familiar habits. EV: electric vehicle; WM: washing machine; TD: tumble dryer; DW: dishwasher.

3.7. The Load Scheduling Returned by the the Prosumer Problem Solution and the Bill Saving

The loads scheduling returned by the prosumer problem solution is reported in Figure 10. With respect to the scheduling due to familiar habits in Figure 3, all these loads are time-shifted according to family members' preferences, i.e., the parameters α_a and β_a. Familiar habits scheduled the recharge of the electric vehicle at midnight; now it is postponed by 2 h. The same delay is applied to the dishwasher. On the contrary, the washing machine and the tumble dryer are relevantly time-shifted; in particular, both these appliances are moved from the late evening to the early afternoon. On 1 July 2015, the householder now pays 0.58 € (it was 0.52 €) for the energy consumption and power losses; on the other hand, he receives 0.28 € (it was 0.15 €) for the exported electricity. The daily net cost now is 0.30 € (it was 0.37 €) therefore a 23.33% bill saving is achieved.

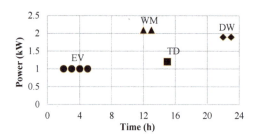

Figure 10. Load scheduling returned by the prosumer problem solution.

3.8. Power at the Point of Delivery after the Prosumer Problem Solution

The power flow profile at the point of delivery as returned by the application of the prosumer problem solution is illustrated in Figure 3 with a continuous line. Although the loads demand and the distributed generators generation remain unchanged, the daily energy exchanged at the point of delivery increases; this is because in a DR program the aggregator has the right to exchange power from the batteries to the grid and vice versa. By comparing the two lines of Figure 3, the electric energy drawn from the grid to the unit increases from 10.17 kWh to 13.6 kWh while the electric energy fed into the grid increases from 4.30 kWh to 6.21 kWh. In addition, Figure 3 clearly shows a marked change in peaks; at 21:00 when the import price and the export price have their maximum value, a fed-power-into-grid peak of about 3 kW substitutes for a drawn-power-from-grid peak of 2.2 kW. Moreover, a new fed-power-into-grid peak of 3 kW now appears at 5:00 a.m.

3.9. State of Charge of Batteries after the Prosumer Problem Solution

The state of charge of batteries (*SOC*), as returned by the application of the prosumer problem solution, is illustrated in Figure 5. Since the aggregator has the right to exchange power between the grid and the batteries, the *SOC* rapidly increases in the early hours of the day because electricity is purchased at a low price. The *SOC* at 5:00 is approximately 85%; a part of stored energy is used to supply loads until 7:00. Two hours later, at 9:00, the *SOC* increases again because of an excess in generation from renewables compared to the load demand; the *SOC* reaches the maximum value at 11:00 and retains this value until about 17:00 when it continuously decreases until the minimum value at 21:00. In order to assess the difference of the use of the batteries between the cases when a DR program is applied and when it is not, we adopt the index:

$$UoB = \sum_{h=1}^{24} |SOC(h) - SOC(h-1)| \qquad (16)$$

where *SOC(h)* is the state of charge at time *h* and the double vertical signs indicate the absolute value of the difference. The calculation returns a *UoB* equal to 1.73 in the case of the DR program and *UoB* equal to 1.45 on the contrary, thus indicating an increase of 19.30%.

3.10. Load Sharing after the Prosumer Problem Solution

When the solution returned by the prosumer problem is applied, the load demand is entirely satisfied by distributed generators and the grid without interruptions. With reference to Figure 8, the grid provides 38.53% (40.28% without the DR program) of load demand while the remaining part is provided by distributed generators. For the latter, 36.10% (21.11% without the DR program) is directly provided by photovoltaic to loads, the wind turbine to loads directly provides 18.74% (20.06% without the DR program), the remaining 6.62% (18.55% without the DR program) is indirectly provided through the batteries.

4. The Laboratory Prototypes of the Proposed Energy Box

This section presents two laboratory prototypes of the proposed energy box, namely low-EB and high-EB; the cost for each prototype is about 100 € or lower. The first prototype low-EB has a limited computing capacity and an Arduino MEGA 2560 (Arduino Holding, Ivrea, Italy) performs it; the second prototype high-EB has a greater computing capacity and a Raspberry Pi3 (Raspberry Pi Foundation, Cambridge, United Kingdom) performs it. Both prototypes are mounted on a demonstration panel of a residential unit together with a real home automation system. A personal computer and Matlab software (R2010a, Mathworks, Natick, MA, USA, 2010) are used to implement the aggregator.

4.1. The Demonstration Panel of a Residential Unit Togheter with a Real Home Automation System

The front and rear sides of the above-mentioned demonstration panel of a residential unit are shown in Figure 11; the demonstration panel is connected to the 230 V/50 Hz utility grid and it is equipped with a real home automation system. The front side of the panel shows the plan of the house. Some icons with LED-lights link to appliances, meters, indoor and outdoor lights; when the LED lights are on, the corresponding peripheral is activated. Nine plugs are placed at the bottom left corner; six of these are smart plugs (see label a in Figure 11) and refer to an equal number of schedulable loads, the remaining three plugs refer to non-controllable loads. The on/off status of smart plugs can be manually set by means of the six smart buttons (see label b in Figure 11) equipped with a KNX interface device model n. MTN670804 (Schneider Electric, Rueil-Malmaison, France).

The rear side of the panel shows the electrical equipment, thermal-magnetic circuit breakers and the home automation system. At the pressure of a smart button, control frames are generated by the button itself and sent via the shielded twisted pair (STP) cable to a switch actuator (see label c in Figure 11) model n. REG-K/8×/16×/10 A; the actuator opens or closes the electrical circuit supplying the corresponding plug and load. A smart energy meter (see label d in Figure 11), model n. REG-k/3 × 230 V, measures the total electrical energy consumption.

Lastly, the label e in Figure 11 indicates a small case (width 240 mm, length 200 mm, depth 100 mm); both the low-EB and high-EB prototypes are mounted inside that case, as better illustrated in the following subsections.

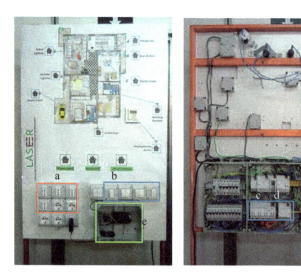

Figure 11. Front and rear side of the demonstration panel mounted with a real home automation. system; **a**: smart plugs; **b**: smart buttons; **c**: switch actuator; **d**: energy meter; **e**: prototype's compartment.

4.2. The Aggregator

The customer coordinates the integrated community, it communicates to customers via the internet by means of the EBs so as to provide services such as the calculation of the optimal scheduling of customers' loads. In the laboratory setup, the aggregator is a software, implemented in Matlab and running on a conventional personal computer. Figure 12 shows the aggregator's graphic-user-interface (GUI) (see label a); the GUI is a conventional web page that the customer visits to upload his preferences regarding the operation of schedulable loads, i.e., the input parameters α_a and β_a of the prosumer problem. The aggregator solves the prosumer problem and calculates the optimal scheduling; Figure 12 also shows the Matlab software and the loads scheduling (see label b).

Figure 12. The aggregator; **a**: graphic-user-interface; **b**: Matlab; **c**: power flow monitoring at the point of delivery; **d**: actual power flow; **e**: forecast power flow.

Furthermore, the aggregator receives and stores the energy consumption measurements returned by EBs; these data allow the aggregator to act as intermediary between the community and the distributed system operator. In particular, the aggregator executes the real-time calculation of power imbalances at the community level and implements strategies to provide ancillary services. With this in mind, Figure 12 (see label c) shows two lines where line d represents the daily forecast of the power flow at the point of delivery of a customer, while the line e represents the actual power flow as returned by the EB of the customer. It is worth noting that in Figure 12, line e stops at the last EB communication, i.e., at 3 p.m.

4.3. The Low-EB Prototype with Arduino

The prototype of the proposed energy box with a limited computing capacity is named low-EB and it mainly consists of: one liquid crystal display (LCD) with four lines and twenty characters per line, one Arduino Mega 2560, one WiFi shield for Arduino, one micro SD card and one sim Tapko KNX (TAPKO Technologies GmbH, Regensburg, Germany). Figure 13 illustrates the low-EB when installed into the case of the demonstration panel of Figure 11. The mission of the low-EB is exclusively to ask the aggregator for the load scheduling, i.e., the prosumer problem solution, and apply the scheduling. More precisely, the low-EB connects to the internet via a local router and synchronizes its internal clock to that provided by the National Institute of Meteorological Research [25]. Then, the low-EB sends a request to the aggregator for the load scheduling. The aggregator calculates the load scheduling in agreement with the customer's preferences, i.e., the parameters α_a and β_a that the customer previously uploaded when he visited the aggregator's GUI (see Figure 11, label a). The load scheduling is delivered to the low-EB in the form of a non-encrypted text file, as illustrated in Figure 14.

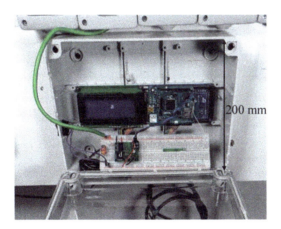

Figure 13. The low-EB prototype and **a** the liquid crystal display (LCD); **b**: the Arduino Mega 2560; c: the WIFI shield; **d**: the sim Tapko and **e** the shielded twisted pair (STP) cable.

The text file starts with the reserved word *BEGIN*; three comments follow (a double slash "//" anticipate each comment). The first comment reports the name of the procedure *psol* that performed the load scheduling and the date/time when the scheduling has been calculated 26_04_2016_10_41; remaining comments report the sender *aggregator* and the recipient *low-EB*. These comments are definitely essential and useful because the low-EB may ask the aggregator for multiple solutions; for instance, the low-EB may ask for "emergency solutions" that is solutions where batteries are out of service or where the customer is suddenly disconnection from the utility grid. Therefore, the low-EB already has a loads scheduling to meet an emergency, pending the aggregator submitting an updated scheduling. Then, the text file reports the loads scheduling in the form of a table with five columns and 24 lines; the first column *INTER* relates to the daily hours from 1 to 23, while the remaining columns refer to the schedulable electrical loads (i.e., the washing machine WM, the tumble drier TD, the dishwasher DW, the electric vehicle EV). Each row is the hourly scheduling and it defines the state on/off of each schedulable load. For instance, the fourth row *INTER* = 4 indicates the time interval from 4:00 a.m. to 4:59 a.m.; for this interval the washing machine is off as the scheduling says "-" as well as the tumble drier and the dishwasher. On the contrary, the recharging of the batteries is running because the scheduling says "*ON*".

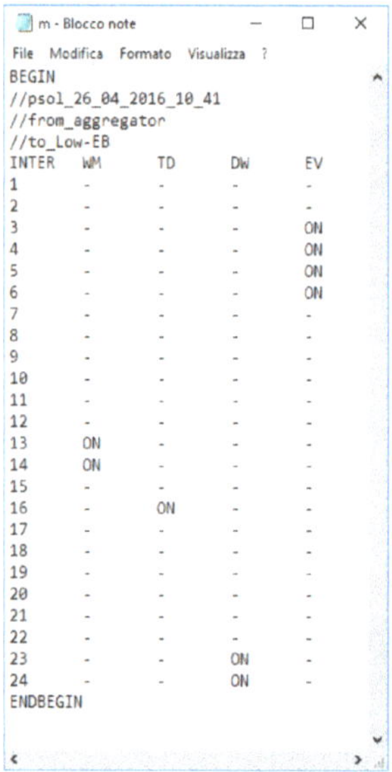


```
BEGIN
//psol_26_04_2016_10_41
//from_aggregator
//to_Low-EB
INTER  WM   TD   DW   EV
1      -    -    -    -
2      -    -    -    -
3      -    -    -    ON
4      -    -    -    ON
5      -    -    -    ON
6      -    -    -    ON
7      -    -    -    -
8      -    -    -    -
9      -    -    -    -
10     -    -    -    -
11     -    -    -    -
12     -    -    -    -
13     ON   -    -    -
14     ON   -    -    -
15     -    -    -    -
16     -    ON   -    -
17     -    -    -    -
18     -    -    -    -
19     -    -    -    -
20     -    -    -    -
21     -    -    -    -
22     -    -    -    -
23     -    -    ON   -
24     -    -    ON   -
ENDBEGIN
```

Figure 14. The optimal load scheduling according to the prosumer problem solution.

At the beginning of every hour, the low-EB reads the corresponding row of the loads scheduling, it generates a set of control frames for each schedulable load, it converts control frames into analog signals and writes them to serial port pins where the sim Tapko KNX is connected. The sim Tapko, in turn, converts analog signals into control frames using the KNX protocol and writes them on the shielded twisted pair (STP) cable of the home automation system. Lastly, every five seconds the low-EB invokes the smart meter for the power measurement at the point of delivery, every minute it calculates the average value of the last 30 values, every 15 min it sends the last 15 mean values to the aggregator.

4.4. The High-EB Prototype Using Raspberry Pi3

The prototype of the proposed energy box with a higher computing capacity is named high-EB and it mainly consists of: a 7 inches touchscreen, one Raspberry Pi3 and sim Tapko KNX. Figure 15 illustrates the High-EB when installed into the case of the demonstration panel of Figure 11. The high-EB performs the same functions performed by the low-EB and, in addition, it is capable to self-calculate the optimal scheduling of consumer's loads. At this scope, the high-EB facilitates the customer in indicating their preferences, α_a and β_a, because the aggregator's GUI (see Figure 12 label a) now locally runs on the Apache web server application installed on the Raspberry Pi3 (see Figure 15 label a). The high-EB communicates with the aggregator and asks for services such as the hourly electricity prices and the hourly PV-wind generation forecast for the customer's site. The high-EB uses the customer's preferences, prices and forecasts to calculate the optimal loads scheduling; the high-EB hourly generates and writes control frames on the STP cable using the sim Tapko KNX in order to apply the loads scheduling.

Figure 15. The high-EB prototype and **a** a touchscreen; **b**: a Rasperry Pi3; **c**: a Sim Tapko and **d** the STP cable.

As well as the low-EB, every five seconds the high-EB carries out a reconnaissance on the system's power consumption; in particular, it invokes the smart meter for the power measurement at the point of delivery and every 15 min it sends the last 15 mean values to the aggregator.

In order to allow the high-EB to self-calculate the loads scheduling, we programmed the prosumer problem into the Raspberry Pi3 using the development environment named Eclipse for Java programming. In order to solve the prosumer problem, we used the IBM ILOG CPLEX Optimization Studio solver (CPLEX); such a solver uses the simplex method, it runs on Linux (i.e., the Raspberry Pi3 operative system) and it solves the prosumer problem within 10–15 s.

5. Conclusions

This paper proposed an energy box (EB) as a viable solution to the challenge of the communication between a consumer and an aggregator; indeed, the proposed EB communicates with the aggregator's IT platform and requires access to an application, service or system. For instance, the proposed EB uploads the user's preferences to obtain optimal loads scheduling and savings on the electricity bill. Furthermore, the proposed EB uploads power/voltage/current measurements to obtain reports and statistics. The proposed EB is also a viable solution to the challenge of the interaction between an EB and home energy management systems (HEMSs); indeed, the proposed EB overcomes the problem of interaction because it directly communicates with all peripherals of the home automation system, bypassing the HEMS. In particular, the proposed EB self-generates the control frames to turn on/off a load or to acquire a measurement from a smart meter; then, the EB sends these frames to loads and smart meters via the physical media for the communication of the home automation system.

The paper also illustrated two prototypes of the proposed EB to have a clear overview of management and material requirements, as well as of cost-effectiveness; both prototypes were tested in the laboratory using a real home automation system. The two prototypes are relevantly cost-effective and effectively serve prosumers and prosumages in autonomous demand response program in cloud-based architectures. The first prototype has a limited calculation capability and a very low cost (low-EB), the second prototype has a higher cost but also a higher capacity for solving calculation problems (high-EB). The low-EB communicates with the aggregator over the internet to exchange data and submit service requests; in particular, the low-EB asks and receives the optimal scheduling of the consumer's loads from the aggregator. The low-EB applies the load scheduling by sending the on/off commands directly to the peripherals of the home automation system, bypassing the HEMS. The high-EB performs the same functions as the low-EB and, in addition, it is capable to self-calculate the optimal scheduling of the consumer's loads. The low-EB uses on an Arduino MEGA 2560, whereas the high-EB uses a Raspberry Pi3.

Future work and research will focus on the relationship between the goals at the aggregator's level and the preferences at the customers' level; at this scope, many field measurements are necessary. With this in mind, we are already involved in a project on power cloud and distributed energy storage systems; we plan to build a certain number of EBs and give them to real customers. In this way, an appropriate quantity of data, information and measurements will be available to evaluate the relationship between goals at the aggregator's and customer's levels.

Acknowledgments: This work was financed by the Italian Ministry of Economic Development (MISE) and the Ministry of Education, University and Research (MIUR), through the National Operational Program for Development and Competitiveness 2007–2013, Project DOMUS PON 03PE_00050_2.

Author Contributions: L.S. and N.S. conceived and designed the numerical experiments; A.B., L.S. and N.S. conceived and designed the laboratory tests; D.M. and L.S. designed the prototypes, L.S. built the prototypes; A.P. and G.B. analyzed the data and contributed analysis tools; A.B. and N.S. wrote the paper.

Conflicts of Interest: The authors declare no conflict of interest.

References

1. Belhomme, R.; Sebastian, M.; Diop, A.; Entem, M.; Bouffard, F.; Valtorta, G. *Deliverable 1.1 Address Technical and Commercial Conceptual Architectures*; Technical Report; Interactive Energy: Lucerne, Switzerland, 2009.
2. Razmara, M.; Bharati, G.R.; Hanover, D.; Shahbakhti, M.; Paudyal, S.; Robinett, R.D., III. Building-to-grid predictive power flow control for demand response and demand flexibility programs. *Appl. Energy* **2017**, *203*, 128–141. [CrossRef]
3. Losi, A.; Mancarella, P.; Vicino, A. *Integration of Demand Response into the Electricity Chain: Challenges, Opportunities and Smart Grid Solutions*, 1st ed.; ISTE Ltd. & John Wiley & Sons: Hoboken, NJ, USA, 2015; ISBN 978-1-84821-854-3.
4. Razzaq, S.; Zafar, R.; Khan, N.A.; Butt, A.R.; Mahmood, A. A Novel Prosumer-Based Energy Sharing and Management (PESM) Approach for Cooperative Demand Side Management (DSM) in Smart Grid. *Appl. Sci.* **2017**, *6*, 275. [CrossRef]
5. Paterakis, N.G.; Pappi, I.N.; Catalao, J.P.; Erdinc, O. Optimal operational and economical coordination strategy for a smart neighborhood. In Proceedings of the 2015 IEEE Eindhoven PowerTech, Eindhoven, The Netherlands, 29 June–2 July 2015; pp. 1–6.
6. Mhanna, S.; Verbič, G.; Chapman, A.C. A faithful distributed mechanism for demand response aggregation. *IEEE Trans. Smart Grid* **2016**, *7*, 1743–1753. [CrossRef]
7. Di Bella, G.; Giarré, L.; Ippolito, M.; Jean-Marie, A.; Neglia, G.; Tinnirello, I. Modeling energy demand aggregators for residential consumers. In Proceedings of the 2013 IEEE 52nd Annual Conference on Decision and Control (CDC), Florence, Italy, 10–13 December 2013; pp. 6280–6285.
8. Ioakimidis, C.S.; Oliveira, L.J. Use of the energy box acting as an ancillary service. In Proceedings of the 2011 8th International Conference on the European Energy Market (EEM), Zagreb, Croatia, 25–27 May 2011; pp. 574–579.
9. Nojavan, S.; Majidi, M.; Esfetanaj, N.N. An efficient cost-reliability optimization model for optimal siting and sizing of energy storage system in a microgrid in the presence of responsible load management. *Energy* **2017**, *139*, 89–97. [CrossRef]
10. Rodriguez-Mondejar, J.A.; Santodomingo, R.; Brown, C. The ADDRESS energy box: Design and implementation. In Proceedings of the 2012 IEEE International Energy Conference and Exhibition (ENERGYCON), Florence, Italy, 9–12 September 2012; pp. 629–634.
11. Donnal, J.S.; Paris, J.; Leeb, S.B. Energy Applications for an Energy Box. *IEEE Internet Things J.* **2016**, *3*, 787–795. [CrossRef]

12. Livengood, D.; Larson, R. The energy box: Locally automated optimal control of residential electricity usage. *Serv. Sci.* **2009**, *1*, 1–16. [CrossRef]
13. Althaher, S.; Mancarella, P.; Mutale, J. Automated demand response from home energy management system under dynamic pricing and power and comfort constraints. *IEEE Trans. Smart Grid* **2015**, *6*, 1874–1883. [CrossRef]
14. Ioakimidis, C.S.; Oliveira, L.J.; Genikomsakis, K.N. Wind power forecasting in a residential location as part of the energy box management decision tool. *IEEE Trans. Ind. Inform.* **2014**, *10*, 2103–2111. [CrossRef]
15. Zhang, D.; Li, S.; Sun, M.; O'Neill, Z. An optimal and learning-based demand response and home energy management system. *IEEE Trans. Smart Grid* **2016**, *7*, 1790–1801. [CrossRef]
16. Marinakis, V.; Doukas, H.; Karakosta, C.; Psarras, J. An integrated system for buildings' energy-efficient automation: Application in the tertiary sector. *Appl. Energy* **2013**, *101*, 6–14. [CrossRef]
17. Cintuglu, M.H.; Youssef, T.; Mohammed, O.A. Development and application of a real-time testbed for multiagent system interoperability: A case study on hierarchical microgrid control. *IEEE Trans. Smart Grid* **2016**, *69*, 1. [CrossRef]
18. Croce, D.; Giuliano, F.; Tinnirello, I.; Galatioto, A.; Bonomolo, M.; Beccali, M.; Zizzo, G. Overgrid: A fully distributed demand response architecture based on overlay networks. *IEEE Trans. Autom. Sci. Eng.* **2017**, *14*, 471–481. [CrossRef]
19. Brusco, G.; Burgio, A.; Menniti, D.; Pinnarelli, A.; Sorrentino, N. Energy management system for an energy district with demand response availability. *IEEE Trans. Smart Grid* **2014**, *5*, 2385–2393. [CrossRef]
20. Brusco, G.; Burgio, A.; Menniti, D.; Pinnarelli, A.; Sorrentino, N. A Power Electronic Device for Controlling A Free-Piston Stirling Engine with Linear Alternator. U.S. Patent 102016000065916, 26 June 2016.
21. Ogwumike, C.; Short, M. Evaluation of a heuristic approach for efficient scheduling of residential smart home appliances. In Proceedings of the 2015 IEEE 15th International Conference on Environment and Electrical Engineering (EEEIC), Rome, Italy, 10–13 June 2015; pp. 2017–2022.
22. Setlhaolo, D.; Xia, X.; Zhang, J. Optimal scheduling of household appliances for demand response. *Electr. Power Syst. Res.* **2014**, *116*, 24–28. [CrossRef]
23. Jovanovic, R.; Bousselham, A.; Bayram, I.S. Residential demand response scheduling with consideration of consumer preferences. *Appl. Sci.* **2016**, *6*, 16. [CrossRef]
24. Autorità per L'energia Elettrica il Gas e il Sistema Idrico. Available online: http://www.autorita.energia.it/elettricita (accessed on 26 April 2016).
25. Istituto Nazionale di Ricerca Metrologica (INRiM). Available online: http://www.inrim.it/ntp/services_i.shtml (accessed on 26 April 2016).

Article

Energy-Efficient Scheduling of Periodic Applications on Safety-Critical Time-Triggered Multiprocessor Systems

Xiaowen Jiang [1] , Kai Huang [1],*, Xiaomeng Zhang [1], Rongjie Yan [2], Ke Wang [1], Dongliang Xiong [1] and Xiaolang Yan [1]

[1]   Institute of VLSI Design, Zhejiang University, Hangzhou 310027, China; xiaowen_jiang@zju.edu.cn (X.J.); xiaomeng_zhang@zju.edu.cn (X.Z.); wangke@vlsi.zju.edu.cn (K.W.); xiongdl@vlsi.zju.edu.cn (D.X.); yan@vlsi.zju.edu.cn (X.Y.)

[2]   State Key Laboratory of Computer Science, Institute of Software, Chinese Academy of Sciences, Beijing 100190, China; yrj@ios.ac.cn

*   Correspondence: huangk@vlsi.zju.edu.cn

Received: 9 May 2018 ; Accepted: 14 June 2018; Published: 19 June 2018

Abstract: Energy optimization for periodic applications running on safety/time-critical time-triggered multiprocessor systems has been studied recently. An interesting feature of the applications on the systems is that some tasks are strictly periodic while others are non-strictly periodic, i.e., the start time interval between any two successive instances of the same task is not fixed as long as task deadlines can be met. Energy-efficient scheduling of such applications on the systems has, however, been rarely investigated. In this paper, we focus on the problem of static scheduling multiple periodic applications consisting of both strictly and non-strictly periodic tasks on safety/time-critical time-triggered multiprocessor systems for energy minimization. The challenge of the problem is that both strictly and non-strictly periodic tasks must be intelligently addressed in scheduling to optimize energy consumption. We introduce a new practical task model to characterize the unique feature of specific tasks, and formulate the energy-efficient scheduling problem based on the model. Then, an improved Mixed Integer Linear Programming (MILP) method is proposed to obtain the optimal scheduling solution by considering strict and non-strict periodicity of the specific tasks. To decrease the high complexity of MILP, we also develop a heuristic algorithm to efficiently find a high-quality solution in reasonable time. Extensive evaluation results demonstrate the proposed MILP and heuristic methods can on average achieve about 14.21% and 13.76% energy-savings respectively compared with existing work.

Keywords: energy; scheduling; multiprocessor systems; safety/time-critical; time-triggered; MILP; heuristic

1. Introduction

Multiprocessor architecture such as Multi-Processor System-on-Chip (MPSoC) are increasingly believed to be the major solution for an embedded cloud computing system due to high computing power and parallelism. The multiprocessor architecture of an MPSoC incorporates multiprocessors and other functional units in a single case on a single die. Meanwhile, there is an ongoing trend that diverse emerging safety-critical real-time applications, such as automotive, computer vision, data collection and control applications are running simultaneously on the MPSoCs [1]. For these safety-related applications, it is imperative that deadlines should be strongly guaranteed. Due to the strong timing requirements and needed predictability guarantees, real-time cloud computing is a complex problem [2]. To satisfy the timing requirement, task scheduling typically relies on an offline schedule based on the architectures such as TTA (Time Triggered Architecture) such that full predictability is guaranteed [3].

In the computing systems, time-triggered scheduling, where tasks have to be executed at particular points in real time, is often utilized to form a deterministic schedule. To effectively schedule the applications, the safety-critical systems require more elaborated scheduling strategies to meet timing constraints and the precedence relationships between tasks.

Reducing energy consumption or conducting green computing is a critical issue in deploying and operating cloud platforms. When the safety-related applications are executed on MPSoCs, reducing energy consumption is another concern since high energy consumption translates to shorter lifetime, higher maintenance costs, more heat dissipation, lower reliability, and, in turn, has a negative impact on real-time performance. Therefore, this growing demand to accommodate multiple applications running periodically on the safety/time-critical multiprocessor systems necessitates to develop an efficient scheduling approach to fully exploit the energy-saving potential. The high energy consumption mainly comes from the energy consumption at the processor level. To reduce power dissipation of processors, Dynamic Voltage and Frequency Scaling (DVFS) and Power Mode Management (PMM (Note that PMM is also referred to as 'Dynamic Power Management (DPM)' [4]. This paper uses the more specific term 'Power Mode Management' to avoid any confusion.)) are two well established system-level energy management techniques. DVFS reduces the dynamic power consumption by dynamically adjusting voltage or frequency while PMM explores idle intervals of a processor and switches the processor to a sleep mode to reduce the static power.

Lots of research has been done on scheduling for energy optimization in real-time multiprocessor embedded systems [5–11]. An application can usually be modeled as a Directed Acyclic Graph (DAG) where nodes denote tasks, and edges represent precedence constraints among tasks. Among these studies, a DAG-based application must be released periodically, whereas tasks in the application can be started aperiodically. In other words, the start time interval between any two consecutive task instances does not need to be fixed to the value of period as long as precedence and deadline constraints can be met. However, such an assumption in these studies is not suitable for the problem of scheduling periodic DAGs on safety/time-critical time-triggered systems. Their solutions in the studies cannot fully guarantee timing predictability and meet the timeliness requirements if the schedulings are not appropriate. Moreover, their methods are merely for single DAG and cannot be directly applied to multiple DAGs.

On the other hand, for a periodic application in time-triggered systems, the scheduling should follow a strictly regular pattern, where besides release time and deadline, the start time of different invocations of a task must be also periodic [12–20]. In this case, most research efforts in energy-efficient scheduling on time-triggered multiprocessor systems, which applied mathematical programming techniques such as Integer Linear Programming (ILP), simply and consistently assume that all tasks in the application are strictly periodic [12–18]. Their time-triggered scheduling approaches are suitable for the tasks that are designed for periodic samplings and actuations.

Nevertheless, tasks within an application are unnecessarily strictly periodic in reality. Today, newly emerging periodic applications may also consist of tasks that do not generate jobs strictly periodically [21]. A typical case can be easily found in the real-world automotive application in engine management system, where most tasks in the application are strictly periodic, and the non-strict periodic tasks are the angle synchronous tasks. Here, for the angle synchronous tasks, the inter-arrival time depend on the revolutions per minute and the number of cylinders of the engine [22,23]. The non-strict periodic tasks are started with relative deadlines corresponding to around an additional 30 degrees of the crankshaft position, after passing a specific rotation of crankshaft position. Therefore, the oversimplified assumption in previous approaches developed for energy-efficient scheduling may impose excessive constraints and degrades scheduling flexibility of the whole system. Furthermore, from the perspective of energy-saving, the excessive constraints result in unnecessary energy consumption. To make readers easy to follow, a motivating example presented in Section 4 will illustrate the problem.

In this paper, we study the energy-efficient scheduling problem arising from the requirements of safety-related real-time applications when deployed in the context of cloud computing embedded platforms. We focus on the problem of static scheduling multiple periodic applications consisting of both strictly and non-strictly periodic tasks on safety/time-critical time-triggered MPSoCs for energy optimization by employing the two powerful techniques: DVFS and PMM. To reduce energy consumption more effectively, both strictly and non-strictly periodic tasks in time-triggered applications should be correctly addressed. This requires an intelligent scheduling that can capture the strict periodicity of specific tasks. Moreover, the problem becomes more challenging when scheduling for energy minimization by combining DVFS with PMM has to consider periodicity of specific tasks in time-triggered systems. In addition, the energy-efficient scheduling problem becomes more complicated as the number of applications running extends from single to multiple. Our main contributions are summarized as the following:

- We consider the unique feature of periodic applications that not all tasks within the applications are strictly periodic in time-triggered systems. A practical task model that can accurately characterize the periodic applications is presented and an energy-efficient scheduling problem based on the model is formulated.
- To solve the problem, we present an improved Mixed Integer Linear Programming (MILP) formulation utilizing the flexibility of non-strictly periodic tasks to reduce unnecessary energy overhead. The MILP method can generate the optimal scheduling solutions.
- To overcome disadvantage of the MILP method when the size of the problem expands, we further develop a heuristic method, named Hybrid List Tabu-Simulated Annealing with Fine-Tune (HLTSA-FT), which integrates the list-based energy-efficient scheduling and tabu-simulated annealing with a fine-tune algorithm. The heuristic can obtain high-quality solutions in a reasonable time.
- We conduct experiments on both synthetic and realistic benchmarks. The experimental results demonstrate the effectiveness of our approach.

It is worth mentioning that, based on the static energy-efficient deterministic schedule (defined in a static configuration file) generated by our proposed methods, the operating system kernel applies it to schedule the partition at its assigned time slot for designing of a practical safety/time-critical partitioned system, where the middleware is integrated to ease interoperability and portability of components to satisfy requirement regarding cost, timeliness, power consumption and so on [24,25].

The remainder of this paper is organized as follows: Section 2 reviews related work in the literature. Section 3 describes models and defines the studied problem. In Section 4, we give a motivating example to explain our idea. Our approach is presented in Section 5. Experimental results are provided in Section 6. The conclusions are presented in Section 7.

2. Related Work

Scheduling for energy optimization is a crucial issue in real-time systems [2,4]. Energy-efficient scheduling of the DAG-based application on the systems have been extensively studied. To name a few, Baskiyar et al. combined DVFS and decisive path scheduling list scheduling algorithm to achieve two objectives of minimizing finish time and energy consumption [6]. Liu et al. distributed the slack time over tasks with the DVFS techniques on the critical path to achieve energy savings [8]. However, they are merely for single DAG-based application. Moreover, these approaches only consider dynamic power consumption, and ignore static power consumption that becomes prominent in the deep submicron domain. In energy-harvesting system, Qiu et al. were devoted to reducing power failures and optimizing the computation and energy efficiency [26], and the authors in [27] addressed the scheduling of implicit deadline periodic tasks on a uniprocessor based on the Earliest Deadline First-As Soon As Possible (EDF-ASAP) algorithm. The works in [28,29] combined DVFS and PMM to minimize energy consumption for scheduling frame-based tasks. However, their approaches can only address independent tasks in a single-processor system. Kanoun et al. proposed a fully self-adaptive energy-efficient online scheduler for general DAG models for multicore DVFS- and PMM-enabled platforms [9]. However, the proposed energy-efficient scheduling solution is designed for soft real-time tasks, where missing deadlines is tolerable.

The aforementioned studies are not applicable for safety-critical applications that have the highest level of safety. Furthermore, in these studies, an application is only periodic in terms of its release time and each task within the application can start aperiodically. Obviously, such an assumption is untenable for a time-triggered application in safety/time-critical systems. The scheduling of tasks in time-triggered systems have also been reported in [13,14,18–20]. Lukasiewycz et al. obtained a schedule for time-triggered distributed automotive systems by a modular framework which provided a symbolic representation used by an ILP solver [13]. Sagstetter et al. studied the problem of synthesizing schedules for the static time-triggered segment for asynchronous scheduling in current automotive architectures, and proposed an ILP approach to obtain optimal solutions and a greedy heuristic to obtain high quality solutions [18]. Freier and Chen presented the time-triggered scheduling policies for real-time periodic task model [19]. Gendy introduced techniques to automate the process of searching for a workable schedule and increase the system predictability [20]. Unfortunately, these works only focus on enhancing system performance.

Research efforts devoted to task scheduling for energy optimization in time-triggered embedded systems have received attention recently. Chen et al. presented ILP formulations and developed two algorithms to address the energy-aware task partitioning and processing unit allocation for periodic real-time tasks [15]. However, the work only addresses independent tasks, and it is not suitable for the DAG-based applications. For periodic dependent tasks, Pop et al. proposed a constraint logic programming-based approach for time-triggered scheduling and voltage scaling for low-power and fault-tolerance [16]. Recently, the state-of-the-art work in [17] introduced a key technique to model the idle interval of the cores by means of MILP. The study proposed a time-triggered scheduling approach to minimize total system energy for a given set of applications represented as DAGs and a mapping of the applications. However, the studies all assume that each task and its instances are started in a strictly periodic pattern. In reality, besides the strictly periodic tasks within time-triggered applications, there also exist non-strictly periodic tasks where each instance of a task does not need to be started periodically [21–23]. To the best of our knowledge, Ref. [21] is the first study that tried to derive better system performance with scheduling both strictly and non-strictly periodic tasks in the safety-critical time-triggered systems. However, their work only focuses on enhancing schedulability, and energy optimization is not involved. In this paper, we address the energy-efficient scheduling problem for periodic time-triggered applications consisting of both strictly and non-strictly periodic tasks.

Methods for scheduling applications on time-triggered multiprocessor systems are mostly based on mathematical programming techniques [12–18]. Since scheduling in multiprocessor systems is NP-hard, many heuristics have been developed when the scale of the problem is increased. To schedule multiple DAG-based applications on real-time multiprocessor systems, the studies in [6,8,21,30–34] presented a collection of static greedy scheduling heuristics based on list-based scheduling. The list-based scheduling heuristics are generally accepted and can provide effective scheduling solutions and its performance is comparable with other algorithms at lower time complexity. They efficiently reduce the search space by means of greedy strategies. However, due to the greedy nature, they can only address certain cases efficiently and cannot ensure the solution quality for a broad range of problems.

On the other hand, to explore the solution space for a high-quality solution, current practice in many domains such as job shop scheduling [35], autonomous power systems [36], distributed scheduling [37] and energy-efficient scheduling problems in embedded system [38–40], favors Tabu Search (TS)/Simulated Annealing (SA) meta-heuristic algorithms. They have shown superiority to the one-shot list scheduling heuristics, despite a higher computational cost. However, both TS and SA have advantages and disadvantages. In general, the SA algorithm is problem-independent, which is analogous to the physical process of annealing. However, it does not keep track of recently visited solutions and needs more iterations to find the best solution. TS algorithm is more efficient in finding the best solution in a given neighborhood, whereas it cannot guarantee convergence and avoid cycling [35]. Moreover, the algorithms cannot be directly used to solve our problem since the non-strictly periodic tasks in time-triggered applications are ignored (whether a task starts strictly periodic or not has a strong influence on scheduling and total energy consumption of the whole system).

To the best of our knowledge, the heuristic method for our problem is not yet reported. In this paper, we consider to solve the problem by formulating the MILP model to obtain optimal solutions, and further to develop an efficient heuristic algorithm since computation time of the MILP method is intolerable when the problem size increases.

3. Problem Formulation

In this section, we first introduce related models and basic concepts that will be used in the later sections, and then provide the problem formulation. The notations used in this paper and their definitions are listed in Table 1.

Table 1. Notations.

Symbol	Description
v_i^m	Task v_i executed on core m ($1 \leq m \leq M$)
$v_{i,p}^m$	The p-th instance of task v_i executed on core m
$Cv_{i,j}$	Communication task (v_i^m transfer data to v_j^n)
$C_{i,j}$	Communication time between v_i^m and v_j^n
$comm(v_i, v_j)$	Communication size between task v_i and v_j
D_k	Deadline of application g_k
P_k	Period of application g_k
d_i	Deadline of task v_i
p_i	Period of task v_i
H_a	Hyper-period of all tasks
B	The bus bandwidth
$W_{i,l}^m$	WCET of task v_i on core m under v/f level l
χ^m	Time overheads for DVFS and task switch on core m
T_{bet}	break-even-time
P_{al}	Total power when the core is active under v/f level l
P_{idle}	Total power when the core is idle
P_{sleep}	Total power when the core is sleep
P_{ba}	Total Power of when the bus is active
P_{bi}	Total power of when the bus is idle
t_{idle}	Idle time interval of the core
t_{sleep}	Sleep time interval of the core
E_{cov}	Total energy overhead of the cores mode switching
E_{comm}	Total communication energy consumption in H_a
E_{comp}	Total computation energy consumption in H_a
$E_{total\_H_a}$	Total system energy consumption in H_a
$P_{total\_H_a}$	Total power of the MPSoC

3.1. System Model

In this paper, we consider a typical MPSoC architecture [17,41,42] shown in Figure 1. The MPSoC architecture consists of M processing cores {core 1, core 2, ..., core M}. Each core has its own local memory, and all cores perform inter-core communication by a high bandwidth shared time-triggered non-preemptive bus to access the main memory. The multi-core platform supports L different voltage or frequency (v/f) levels and a set of total power value {P_{a1}, P_{a2}, ..., P_{aL}} ($P_{a1} > P_{a2} > \ldots > P_{aL}$) corresponding to v/f levels. The bus controller implements a given bus protocol (e.g., time-division multiple access protocol), and assigns bus access rights to individual cores. The communication procedure among inter-core rely on message-passing [30]. The characteristics of the system model are as follows: (1) a DVFS- and PMM-enabled MPSoC; (2) non-preemptive; (3) shared time-triggered bus based on a given protocol; (4) communications are supposed to perform at the same speed without contentions; and (5) each core has independent I/O unit that allows for communication and computation to be performed simultaneously. Note that the real communication cost occurs only in inter-core communications where dependent tasks mapped on different cores. In addition, when the tasks are allocated to the same core, the communication cost becomes zero as the intra-core communication can be ignored.

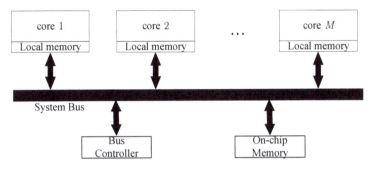

Figure 1. System model.

3.2. Task Model

An application can be modeled by a DAG (or called task graph) comprising a set of dependent nodes connected by edges. This article assumes a periodic real-time task model in which $G = \{g_1, g_2, \ldots, g_K\}$ is a set of K applications to be executed on the MPSoC. Application $g_k \in G$ is denoted as $g_k = \{V_k, E_k, D_k, P_k\}$, where V_k and E_k are set of nodes and edges in g_k, respectively, and D_k and P_k are deadline and period of g_k, respectively. The deadline D_k is assumed to be a constrained deadline, i.e., it is less than or equal to the period P_k. Tasks in g_k share the same period and deadline of g_k. We use H_a to describe the least common multiple of the periods of all tasks, which is called the hyper-period. It is well known that scheduling in a hyper-period gives a valid schedule [43].

In a task graph, each node $v_i \in V_k$ denotes a computation task and each edge $e_j \in E_k$ represents a communication task. The computation tasks complete data computation on the processing cores, and the communication tasks assigned to the bus complete data transmission between the cores. Computation and communication tasks can be performed in parallel since the communication operation is non-blocked. The weighted value on the edge indicates the amount of data transferred between connected computation tasks. The worst case execution time (WCET) of a task v_i on a core m under v/f level l is denoted by $W_{i,l}^m$. These profiling information of tasks can be obtained in advance.

On the MPSoC platform, we consider multiple time-triggered applications which are released periodically. As not all tasks within the applications started strictly periodically, we analyze characteristics of the tasks and make a classification of tasks in an application as follows:

1. Strictly periodic task: the task in an application should strictly start its instances periodically, which means that the start time interval between two successive task instances is fixed. As a strictly periodic task v_1 shows in Figure 2a, in addition to release time and deadline for the application, the start time of different invocations of the task need to also be periodic.
2. Non-strictly periodic task: the task in an application need not start its instances periodically, i.e., the start time interval between two successive task instances is not fixed. As a non-strictly periodic task v_2 shows in Figure 2b, the start time of different invocations of the task can be aperiodic as long as the deadline of the task can be guaranteed.

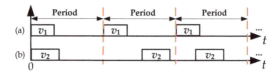

Figure 2. Strictly and non-strictly periodic tasks.

In this paper, according to the existence of the non-strictly periodic task, an application is regarded as exactly periodic if all tasks within the application are strictly periodic; otherwise, it is regarded as a loose periodic application.

3.3. Energy Model

Assuming that the MPSoC supports both DVFS and PMM. In this paper, we adopt the same energy model due to its generality and practicality [17,42,44,45]. The total system energy consumption is composed of energy overhead of communication and computation. We assume each processing core has three modes, active, idle and sleep mode, and the shared bus has two modes, active and idle mode. Various practical issues including time and energy overhead of the core mode switching and inter-core communications are also considered in the energy model. We apply inter-task DVFS [5,17,41] technique, where the supply v/f of the core cannot be changed within single task. The dynamic power consumption of a processing core and operation frequency f are given by:

$$P_d = C_{ef} \times V_{dd} \times f, \tag{1}$$

$$f = k \times (V_{dd} - V_t)^2 / V_{dd}, \tag{2}$$

where C_{ef} is the effective switching capacitance, V_{dd} is the supply voltage, k is a circuit dependent constant and V_t is the threshold voltage. The static power consumption, P_s, is given by:

$$P_s = V_{dd} \times I_{leak}, \tag{3}$$

where I_{leak} denotes leakage current. Therefore, total power consumption when the core is active under v/f level l can be computed as:

$$P_{al} = P_d + P_s + P_{on}, \tag{4}$$

where P_{on} is the intrinsic power that is needed to keep the core on. Thus, the energy consumption of task v_i^m executed on core m at v/f level l can be represented as:

$$E_{al} = W_{i,l}^m \times P_{al}. \tag{5}$$

When a core does not execute any task (idle mode), its power consumption is primarily determined by the idle power. We assume that P_{idle} and P_{sleep} respectively represent the idle power and sleep power. Normally, we have $P_{idle} > P_{sleep}$. Considering the overhead of switching the processing core between active mode and sleep mode, the definition of break-even time T_{bet} is defined as the minimum time interval for which entering the sleep mode is more effective (energy-wise) when compared to the idle mode, despite of an extra time and energy overhead associated to the mode switch between active mode and sleep mode. In other words, the core should keep in idle mode if the idle interval $t_{idle} < T_{bet}$; otherwise, the core should enter into sleep with power consumption P_{sleep}. Similar to [17], T_{bet} can be calculated as:

$$T_{bet} = \max\{t_{ms}, (E_{ms} - P_{sleep} \times t_{ms})/(P_{idle} - P_{sleep})\}, \tag{6}$$

where t_{ms} and E_{ms} are time and energy overhead of the core mode switching, respectively. The energy consumed in idle mode (E_{idle}) and sleep mode (E_{sleep}), are calculated respectively as follows:

$$E_{idle} = P_{idle} \times t_{idle}, \tag{7}$$

$$E_{sleep} = P_{sleep} \times t_{sleep}. \tag{8}$$

Therefore, given a static time-triggered schedule S, the total energy consumption of the processing core is:

$$E_{comp}(S) = E_{al}(S) + E_{idle}(S) + E_{sleep}(S) + E_{cov}(S). \tag{9}$$

The processing core, the bus, and the shared on-chip memory in the architecture complete the data transfer between the two dependent tasks. Specifically, an inter-core communication is issued when two tasks with data dependence are mapped to different processing cores. In addition, the shared on-chip memory stores the intermediate communication. The processing core can initiate a write operation to the shared on-chip memory by providing an address with control information that typically requires one bus clock cycle. The communication time overhead (or latency) refers to the length of time that a message containing multiple words delivered from a source processing core to a target processing core. In the architecture, only one component (e.g., processing core) is allowed to use the bus actively at any one time according to the characteristics of the shared bus. The communication procedure on the shared bus is non-interruptible, thus multiple communications should be serialized. The communication time overhead is proportional to the data transfer size, i.e., $C_{i,j} = comm(v_i, v_j) / B$, where $comm(v_i, v_j)$ is the amount of data transferred between task v_i and task v_j, and B is the communication bandwidth [41]. On chip memory will allocate memory space to store intermediate data. The required memory space will be released until the target processing core sends back to the bus controller the successful data transfer. For a task graph, there is no inter-core communication if both the source and the target node of an edge are mapped on the same core. The inter-core communication energy overhead between task v_i and v_j is calculated as $E_{comm}(v_i, v_j) = C_{i,j} \times P_{ba}$, where P_{ba} is the power of active bus.

3.4. Problem Statement

Mapping and scheduling in multiprocessor systems have each been proven to be NP-hard. In this paper, we decouple the problem into mapping and scheduling. It is worth mentioning that we assume the task mapping can be performed by using any algorithms in previous excellent works [10,30,31]. The energy-efficient scheduling problem is defined as illustrated in Figure 3.

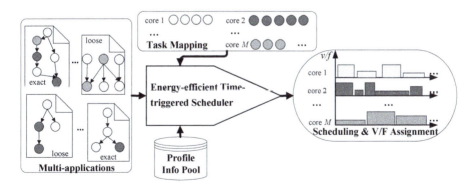

Figure 3. Energy-efficient time-triggered scheduling problem for strict and non-strict periodic tasks.

Given a DVFS- and PMM-enabled MPSoC shown in Figure 1, multiple periodic applications consisting of both strictly and non-strictly periodic tasks, task mapping and profiling information as inputs, the energy-efficient time-triggered scheduler is to find a static non-preemptive scheduling and a v/f assignment for each task in a hyper-period H_a such that total system energy consumption $E_{total\_H_a}$ is minimized while timing constraints are guaranteed.

4. Motivating Example

For easy understanding, in this section, we first present a motivating example to show that state-of-the-art energy-efficient scheduling on time-triggered systems may not work well on the problem. Assuming that an MPSoC has CORE1 and CORE2, each with a high frequency level f_H and a low frequency level f_L. The total active power of the core under f_H and f_L is denoted by P_{aH} and P_{aL}. Assuming a set of applications (denoted as g_1, g_2 and g_3) and their task mappings on the MPSoC have been given as illustrated in Figure 4. g_1 is an exact periodic application in which task v_1, v_2, v_3 and v_4 are responsible for collecting data from sensors periodically and g_2 is a loose periodic application in which task v_5, v_6, v_7 and v_8 are responsible for performing processing data. The edges e_1, e_4, e_6 and e_7 indicate their connected tasks are mapped to different cores and the dashed edges e_2, e_3 and e_5 indicate the corresponding tasks are mapped to the same core. Thus, $comm(v_1, v_3)$, $comm(v_2, v_4)$ and $comm(v_5, v_8)$ are equal to 0. The periods for g_1, g_2 and g_3 are 60, 30 and 60, respectively. Task WCETs and power profiles are shown in Table 2. For simplicity, time unit is 1 ms, power unit is 1 W, and the energy unit is 1 mJ.

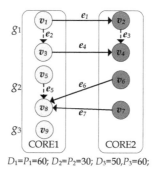

$D_1=P_1=60$; $D_2=P_2=30$; $D_3=50,P_3=60$;

Figure 4. Task graphs and given mapping.

Table 2. Task WCETs and power profiles.

v_i	WCET(f_H)	WCET(f_L)	e_i	$C_{i,j}$
v_1	8	16	e_1	8
v_2	5	10	e_2	4
v_3	7	14	e_3	5
v_4	3	6	e_4	5
v_5	2	4	e_5	6
v_6	6	12	e_6	3
v_7	7	14	e_7	9
v_8	4	8	–	–
v_9	13	26	–	–
$P_{aH} = 0.68$	$P_{aL} = 0.41$	$P_{idle} = 0.19$	$P_{ba} = 0.1$	
$P_{sleep} = 0$	$P_{bi} = 0$	$T_{bet} = 18$	$E_{cov} = 0.6$	

The hyper-period H_a is 60 if we schedule g_1 and g_2. In one hyper-period, g_1 and g_2 are released 1 and 2 times, respectively, as well as each task within its application. Based on the assumptions that the start time interval of any two successive instances of a task must be fixed in previous works [15–17], the scheduling for energy minimization is shown in Figure 5a. In the schedule, the horizontal axis represents the time, and the heights of task blocks represent the frequency level. The start time interval between two consecutive task instances $(v_{5,2}^2, v_{5,1}^2)$, $(v_{6,2}^1, v_{6,1}^1)$, $(v_{7,2}^1, v_{7,1}^1)$ and $(v_{8,2}^2, v_{8,1}^2)$ in g_2 are equal to 30. The average power consumption in H_a can be calculated as 0.854 W.

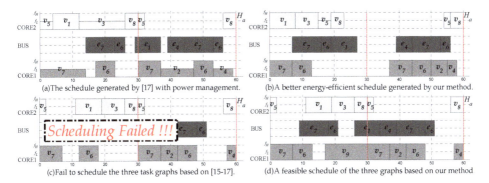

(a) The schedule generated by [17] with power management.

(b) A better energy-efficient schedule generated by our method.

(c) Fail to schedule the three task graphs based on [15–17].

(d) A feasible schedule of the three graphs based on our method

Figure 5. A motivating example.

In contrast to the scheduling from Figure 5a, the schedule generated by our method in Figure 5b shows a better scheduling for energy-efficiency that the average power consumption in H_a is 0.784 W. In the schedule, the start time interval between task pair $(v_{5,2}^2, v_{5,1}^2)$, $(v_{6,2}^1, v_{6,1}^1)$, $(v_{7,2}^1, v_{7,1}^1)$ and $(v_{8,2}^2, v_{8,1}^2)$ in g_2 do not have such the strict constraint of periodicity, as g_2 is a loose periodic application. Due to the flexibility of those non-strictly periodic tasks in the scheduling, increase of total core sleep time and decrease of energy overhead from mode switching can achieve about 8.2% total energy savings compared with Figure 5a.

Assuming that g_3 is a exact periodic application in which v_9 is responsible for data transformation, we then consider to schedule the three task graphs, g_1, g_2 and g_3. The hyper-period is still 60. We find that scheduling the task graphs based on the simplistic assumptions in [15–17] would fail as shown in Figure 5c, as their methods impose overly strict constraints on the task instances within g_2. For example, the start time of the two task instances $v_{6,1}^1$ and $v_{6,2}^1$ are 12 and 42, respectively, such that the start time interval between these two task instances is fixed as 30. Thus, v_9 cannot be scheduled in a hyper period since the size of v_6 is 6 and the size of v_9 is 13. However, there actually exists a feasible schedule as shown in Figure 5d. For the non-strictly periodic task v_6 in g_2, the constraint regarding the periodic interval between the start time of $v_{6,1}^1$ and its next instance $v_{6,2}^1$ is unnecessary. In the schedule, the two task instances, $v_{6,1}^1$ and $v_{6,2}^1$ start at 7 and 42, respectively, and thus v_9 can be scheduled at 17.

From the above results, one can observe that the previous studies which do not consider characteristics of non-strictly periodic tasks may result in more energy consumption and even degradation of the schedulability of the whole system. For our problem in this paper, to reduce energy consumption more effectively, a scheduling approach which is aware of the periodicity of specific tasks and utilizes the flexibility of non-strictly periodic tasks is desired.

5. The Proposed Methods

This section presents the energy-efficient scheduling approach jointly with DVFS and PMM techniques for multiple periodic applications consisting of strictly and non-strictly periodic tasks. With consideration of strictness of tasks' periodicity, we formulate a MILP model to solve the problem and to obtain an optimal scheduling in which the system total energy consumption is the minimum. Then, we develop a heuristic algorithm when the MILP formulation cannot be used to efficiently solve large scale instances.

5.1. MILP Method

ILP-based methods have the advantages of reachable optimality and easy access to various solving tools. We aim to find an energy-efficient time-triggered scheduling and a v/f assignment of all tasks in given an MPSoC, multiple DAGs and task mapping, such that the total system energy consumption is minimized under timing constraints. To obtain optimal energy-efficient scheduling solutions for pre-mapped tasks with consideration of strictness of tasks' periodicity, we now develop our MILP formulation for the problem based on the practical models defined in Section 3. We build up our MILP formulation step by step, including v/f selection constraints, deadline constraints, periodicity constraints for the strictly periodic tasks, precedence constraints, non-preemption constraints, and an objective function. Firstly, we define the following variables:

$x_{i,p,l}^m$: binary variable, $x_{i,p,l}^m = 1$ if v/f level of task $v_{i,p}^m$ is l and $x_{i,p,l}^m = 0$ otherwise,
$ts_{i,p}^m$: Start time of computation task $v_{i,p}^m$,
$cts_{i,p;j,q}^{m;n}$: Start time of communication task $Cv_{i,p;j,q}^{m;n}$ ($v_{i,p}^m$ transfer data to $v_{j,q}^n$),
$O_{s,p;t,q}^m$: binary variable, $O_{s,p;t,q}^m = 1$ if task $v_{s,p}^m$ executes before task $v_{t,q}^m$ and $O_{s,p;t,q}^m = 0$, otherwise.

Then, given task graphs and task mappings, we formulate the MILP model as follows:
Minimize:

$$E_{total\_H_a} = E_{comp} + E_{comm}, \tag{10}$$

Subject to

1. Voltage or frequency selection constraints for each task as we use inter-task DVFS:

$$\Sigma_{l=1}^{L} x_{i,p,l}^m = 1. \tag{11}$$

2. Deadline constraints (χ^m denotes time overheads for DVFS and task switch on core m):

$$ts_{i,p}^m + \Sigma_{l=1}^{L} (x_{i,p,l}^m \times W_{i,l}^m) + \chi^m \leq d_i. \tag{12}$$

3. According to strictness of the tasks's periodicity, we separately determine the start time of these tasks and their instances in H_a. Therefore, for the strictly periodic tasks belonging to the time-triggered applications within one hyper-period H_a, the periodic constraints can be represented as follows:

$$ts_{i,p+1}^m - ts_{i,p}^m = p_i. \tag{13}$$

For any non-strictly periodic task and its instances in H_a, the periodic constraint is unnecessary, that is, the interval between the start time of two consecutive instances of the task is no longer fixed as p_i.

4. Dependency constraints for computation tasks (e.g., source task $v_{i,p}^m$ and target task $v_{k,r}^m$ mapped to the same core:

$$ts_{i,p}^m + \Sigma_{l=1}^{L} (x_{i,p,l}^m \times W_{i,l}^m) + \chi^m \leq ts_{k,r}^m. \tag{14}$$

5. Dependency constraints for tasks (e.g., source task $v_{i,p}^m$ and target task $v_{j,q}^n$ mapped to different cores.

    (a) $Cv_{i,p;j,q}^{m;n}$ can be started only after $v_{i,p}^m$ completes:

$$ts_{i,p}^m + \Sigma_{l=1}^{L} (x_{i,p,l}^m \times W_{i,l}^m) + \chi^m \leq cts_{i,p;j,q}^{m;n}, \tag{15}$$

    (b) $v_{j,q}^n$ can be started only after $Cv_{i,p;j,q}^{m;n}$ completes:

$$cts_{i,p;j,q}^{m;n} + C_{i,j} + \chi^m \leq ts_{j,q}^n. \tag{16}$$

6. Any two computation task instances mapped to the same core must not overlap in time, as well as the communication tasks in the bus. They can only be executed sequentially. Assume task $v_{s,p}^m$

and task $v_{t,q}^m$ are two task instances, and MAX is a constant far greater than H_a. To guarantee either task $v_{t,q}^m$ can run after task $v_{s,p}^m$ finishes, or vice versa, the non-preemption constraint can be expressed as follows:

$$ts_{s,p}^m + \Sigma_{l=1}^L (x_{s,p,l}^m \times W_{s,l}^m) + \chi^m \le ts_{t,q}^m + MAX \times (1 - O_{s,p;t,q}^m), \tag{17}$$

$$ts_{t,q}^m + \Sigma_{l=1}^L (x_{t,p,l}^m \times W_{t,l}^m) + \chi^m \le ts_{s,p}^m + MAX \times O_{s,p;t,q}^m. \tag{18}$$

The two formulas are also applicable to communication tasks mapped to the bus. The differences between computation and communication tasks are that execution time of computation tasks are variable, while communication time of communication tasks are constant. We can get real computational time of the task on the specified core and real communication cost between the dependent tasks, as task mapping has been given. In addition, communication tasks can be overlapped with the computation tasks independent on them.

To formulate the time interval (int_i) of any two adjacent tasks on each core m in one hyper-period H_a, we use the interval modelling technique in [17]. The readers interested in the detailed steps of modeling can refer to [17]. Then, according to the definition of T_{bet}, for each time interval int_i on the core, we have

$$d_i = \begin{cases} 1, & if\,int_i \ge T_{bet}, \\ 0, & if\,int_i < T_{bet}, \end{cases} \tag{19}$$

where d_i refers to a binary variable in the decision array $darray[N]$, representing whether the core should remain idle mode ($d_i = 0$) or enter into sleep mode ($d_i = 1$). Assuming there are N tasks on the core in H_a, the total idle and sleep interval (t_{idle} and t_{sleep}) can be represented as follows:

$$t_{idle} = \Sigma_{1 \le i \le N}((1 - d_i) \times int_i), \tag{20}$$

$$t_{sleep} = \Sigma_{1 \le i \le N}(d_i \times int_i). \tag{21}$$

The total energy overheads of mode switch for the core can be calculated as follows:

$$E_{cov} = \Sigma_{1 \le i \le N}(d_i \times E_{ms}). \tag{22}$$

Note that the step function introduced by d_i in Equation (19) and the multiplication of int_i and binary variable d_i in Equations (20) and (21) are nonlinear equations. Such problems can be solved by commercial or open-source ILP solvers after linearization. Solutions to similar problems have been presented in [46]. We now present the linearization process for our problem.

To linearize the multiplication of $d_i \times int_i$, we define a new variable r_i, such that $r_i = d_i \times int_i$. It is obvious that $int_i \le H_a$. The multiplication can be linearized as the following constraints:

$$r_i - int_i \le 0, \tag{23}$$

$$r_i - d_i \times H_a \le 0, \tag{24}$$

$$int_i - r_i + d_i \times H_a \le H_a. \tag{25}$$

The step function introduced by d_i in Equation (19) can be transformed to the following constraint:

$$d_i \times (int_i - T_{bet}) + (1 - d_i) \times (T_{bet} - int_i) \ge 0. \tag{26}$$

In Equation (26), the multiplication of $d_i \times int_i$ are linearized by using Equations (23)–(25).

Based on these formulations, lastly, we can obtain an optimal scheduling and minimum overall energy consumption $E_{total\_H_a}$ by solving the MILP model with ILP solver.

Limitation of the MILP-based method: Though we can obtain an optimal solution by solving the MILP formulation with modern ILP solvers, it is time-consuming to search the optimal solution for our problem. Specifically, to construct time interval for each task instance in one hyper-period, the time interval modeling in the previously discussed MILP-based method particularly yields a large number of variables, and results in dramatically increased exploration space. The problem may even not be solved because of memory overflow when input size of tasks to be scheduled is large. To address this, we propose an efficient heuristic algorithm to reduce the exponentially increasing scale in Section 5.2.

5.2. Heuristic Algorithm

In this section, we develop a heuristic algorithm, named Hybrid List Tabu-Simulated Annealing with Fine-Tune (HLTSA-FT). Different from the TS/SA algorithm mentioned in Section 2, the proposed algorithm has the following innovations: (1) the HLTSA-FT integrates list scheduling with TS/SA to take advantage of both algorithms and to mitigate their adverse effects. Based on our problem, the decomposition and solution process is iteratively guided by the HLTSA-FT algorithm that employs the proper intensification and diversification mechanism. In HLTSA-FT, the SA supplemented with a tabu list can reduce the number of revisiting old solutions and cover a broader range of solutions; (2) list-based scheduling performed in the List-based and Periodicity-aware Energy-efficient Scheduling (LPES) function can efficiently obtain feasible solutions for our problem; (3) in addition, solutions can be further improved by applying problem specific and heuristic information to guide the process of optimization. Specifically, a fine-tune phase performed in the *FT* function is presented to make minor adjustments of the accepted solution to find a better solution more rapidly. Therefore, the total number of iterations can be reduced and solution quality can be improved. The details of HLTSA-FT are given in Algorithm 1. Three main steps in the algorithm are:

Initialization. The step (in Lines 1–3) first sets appropriate parameters including the initial temperature T_0, the maximum number of iterations $LPMAX$, the maximum number of consecutive rejections $RMAX$, the cooling factor δ and the maximum length of the tabu list TL. Then, the algorithm builds TL with length of $TLIST\_LEN$, and sets aspiration criterion A. The *initial_solution_gen()* function generates an initial solution λ_0 (v/f assignment for each task instance in H_a) as the starting point of optimization process. Since a good initial solution can accelerate the convergence process, the function integrates the MILP model of a relaxed formulation (e.g., by neglecting the idle and sleep interval formulations). λ_0 is evaluated and the current optimal energy consumption is denoted as E_{cur}. The aspiration criterion accepts the move provided that its total energy is lower than that of the best solution found so far. It helps with restricting the search from being trapped at a solution surrounded by tabu neighbors. The tabu list stores recently visited solutions and helps saving considerable computation time by avoiding revisits.

Iteration. In each iteration (in Lines 5–26), the *solution_neighbor()* function (Line 5) generates neighborhood λ_{new} by applying a small perturbation (swap move) to current solution λ_{cur}. In our context, v/f assignments for the tasks are in a neighbourhood. λ_{new} is generated in two steps: (i) select two tasks; and (ii) swap their v/f levels. Then, λ_{new} is checked for feasibility (i.e., if constraints mentioned above are met) by *solution_feasible* function. If the solution is not in the tabu list or satisfies the aspiration criterion, it is selected. Otherwise, a new solution is regenerated. Then, the solution is translated to an energy-efficient schedule by using the *LPES* function (Line 6). The solution λ_{new} which consumes less energy will always be accepted, and when λ_{new} is an inferior solution, it may still be accepted with a probability $Pro(\Delta E, T) = \exp(-\Delta/T)$ where T is the annealing temperature at current iteration. This transition probability can help the algorithm to escape from local optima. Once accepted, λ_{new} is put in the tabu list TL and the current solution is updated by replacing λ_{cur} with λ_{new} for next iteration. Then, the FT phase (Line 16) will performed to fine-tune the accepted λ_{new}. Otherwise, the solution λ_{new} is discarded with *rjnum* plus 1. The algorithm then decreases the temperature and continues to the next iteration.

Algorithm 1: The HLTSA-FT Heuristic Algorithm

Input: DAGs, task mappings and profiles, power profiles

Output: An energy-efficient task schedule S, v/f assignments

1 Set appropriate value of T_0, $LPMAX$, $RMAX$, δ, $TLIST\_LEN$;

2 $T \leftarrow T_0$, $rjnum$, $loop \leftarrow 0$, build tabu_list $TL \leftarrow \Theta$, $\lambda_0 \leftarrow initial\_solution\_gen()$, $\lambda_{opt} \leftarrow \lambda_0$

   $\lambda_{cur} \leftarrow \lambda_0$;

3 $E_0 \leftarrow LPES(\lambda_0)$, $E_{opt} \leftarrow E_0$, $E_{cur} \leftarrow E_0$;

4 **while** *loop < LPMAX and rjnum < RMAX* **do**

5      $\lambda_{new} = solution\_neighbor(\lambda_{cur})$;

6      $E_{new} = LPES(\lambda_{new})$;

7   **if** *solution_feasible and* $(\lambda_{new} \notin TL$ *or* $(\lambda_{new} \in TL$ *and* $E_{new} > A))$ **then**

8 | goto line 13;

9   **end**

10   **else**

11 | goto line 5;

12   **end**

13     $\Delta E \leftarrow (E_{new} - E_{cur})$;

14   **if** $\Delta E < 0$ **then**

15         $\lambda_{cur} \leftarrow \lambda_{new}$, $E_{cur} \leftarrow E_{new}$, $\lambda_{new} \in TL$, $A \leftarrow E_{new}$;

16         $\lambda_{fine} = FT(\lambda_{new})$, $rjunm \leftarrow 0$;

17   **end**

18   **else**

19   **if** $Pro(\Delta E, T) > random()$ *and* $Pro(\Delta E, T) < 1$ **then**

20 | $\lambda_{cur} \leftarrow \lambda_{new}$, $E_{cur} \leftarrow E_{new}$, $\lambda_{new} \in TL$, $A \leftarrow E_{new}$;

21   **end**

22   **else**

23 | $rjnum \leftarrow rjnum + 1$;

24   **end**

25   **end**

26     $loop \leftarrow loop + 1$, $T \leftarrow T \times \delta$;

27 **end**

28 **return** λ_{opt}, E_{opt};

Stopping Criteria. The search procedure will be stopped if the number of iterations or the variable *rjnum* reaches the predefined value. The variable *rjnum* stores the current number of continuous rejections, and it represents that no superior solution exists in the neighborhood and the search has reached a near optimal solution once *rjnum* reaches *RMAX*.

In the next two subsections, we give a detailed description of the *LPES* and *FT*.

5.2.1. LPES

To obtain a feasible scheduling for energy reduction efficiently, the scheduling for our problem needs to addresses two aspects. First, a priority assignment (i.e., execution order of tasks) must satisfy the corresponding constraints (including deadline and precedence constraints for each task graph, and periodicity constraints for the strictly periodic tasks) in the schedule, and maximize the total interval available for energy management. Second, the intervals need to be allocated efficiently to reduce energy consumption. In this paper, we apply the List-based and Periodicity-aware Energy-efficient Scheduling (LPES) method. The first aspect is addressed through bottom level (b-level) based priority assignment. The second aspect is addressed through a modified simple MILP model whose number of variables is only linear with the number of tasks.

List scheduling is a type of scheduling heuristics in which ready tasks are assigned priorities and ordered in a descending order of priority. A ready task is a task whose predecessors have finished executing. Each time, the task with the highest priority is selected for scheduling. If more than one task has the same priority, ties are broken using the strategy such as the random selection method. Priority assignment based on the b-level has been adopted in energy-aware multiprocessor scheduling. The b-level of a task is defined as the length of the longest path from the beginning of the task to the bottom of the task graph. As we focus on multi-DAGs in our problem, we define the b-level of task $v_{i,p}^m$ and its next instance $v_{i,p+1}^m$ within H_a as:

$$BL(v_{i,p}^m) = BL(v_{i,p+1}^m) + p_i. \tag{27}$$

We calculate b-level values of all tasks and their instances, and sort them in a list which is ordered in descending order. The higher the value of b-level, the higher the priority of the task. An example is shown as below:

Example 1. *Consider the case given in the form of two task graphs g_1 and g_2 in Figure 4. The b-level of tasks in H_a are shown in Table 3. Thus, execution order of tasks on CORE1 and CORE2 are denoted as $\{v_{7,1}^1 \to v_{6,1}^1 \to v_{7,2}^1 \to v_{2,1}^1 \to v_{6,2}^1 \to v_{4,1}^1\}$ and $\{v_{5,1}^2 \to v_{8,1}^2 \to v_{1,1}^2 \to v_{3,1}^2 \to v_{5,2}^2 \to v_{8,2}^2\}$, respectively.*

Table 3. The b-level of each task in one hyper-period.

Tasks	$v_{1,1}^2$	$v_{2,1}^1$	$v_{3,1}^2$	$v_{4,1}^1$	$v_{5,1}^2$	$v_{6,1}^1$	$v_{7,1}^1$	$v_{8,1}^2$	$v_{5,2}^2$	$v_{6,2}^1$	$v_{7,2}^1$	$v_{8,2}^2$
b-level	29	13	15	3	42	43	50	34	12	13	20	4

Based on the given priority and v/f assignment for each task, a scheduling with PMM should be generated to reduce total energy consumption. The time interval can be directly modeled as the following:

Assuming that there are N task instances on a core m in a hyper-period H_a, all these tasks are stored in a task list represented as $T_1, T_2, \ldots, T_i, \ldots, T_N (1 \le i \le N)$ where tasks are ordered in descending order of their priorities. As the tasks in the first hyper-period shown in Figure 6, the time interval between any two adjacent tasks, T_i and $T_{i+1}(1 \le i \le N-1)$, can be directly calculated as:

$$int_i = st(T_{i+1}) - ft(T_i), \tag{28}$$

where $st(T_i)$ and $ft(T_i)$ denote start time and finish time of task T_i, respectively. As we focus on task scheduling in one hyper-period and task execution are repeated in each hyper-period, there are N time intervals. The last time interval int_N between task T_1 and task T_N is calculated as $int_N = [H_a - ft(T_N)] + [st(T_1) - 0]$.

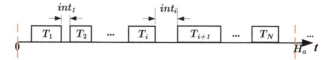

Figure 6. Tasks in a hyper-period.

In the time interval modeling in Section 5.1, a large number of intermediate integer variables are used to check timing information of every task instance to determine the closest task instance for a task instance. While compared with the time interval modeling in [17], the number of constraints and many decision variables (e.g., $x_{i,p,l}^m$ and $O_{s,p;t,q}^m$) and the intermediate variables (e.g., $A_{s,p;t,q}^m$, $B_{s,p;t,q}^m$ and $O_{s,p;t,q}^m - B_{s,p;t,q}^m$ in [17]) have been greatly reduced. After obtaining each idle time int_i, we use the ILP solver to obtain an energy-efficient scheduling.

5.2.2. FT

To find solutions that can further reduce energy consumption, a fine-grained adjustment of the neighborhood range is performed for the accepted solution (line 16 in Algorithm 1). We now present the details of FT phase. The core idea of FT is to increase the potential energy savings by tuning priorities that still satisfy corresponding constraints of task graphs (not blindly or randomly adjusting priorities). In this study, for the strictly periodic tasks, their priorities remain unchanged since the strictness of periodicity of task start time limits the possibility of adjustment within a hyper-period. The non-strictly periodic task instances in H_a do not have to follow the strict condition that all tasks need to be started periodically; thus, they have space for execution-order adjustment. On the other hand, the tasks that have the same b-level value (tie-breaking tasks) also have chances to adjust their priorities. We focus on these tasks that may have better schedule flexibility and, correspondingly, make full use of it to achieve more energy savings. The pseudo-code of the FT is listed in Algorithm 2.

Algorithm 2: The FT Algorithm

Input: DAGs, task mappings, task and power profiles, priority assignments
Output: An improved energy-efficient task schedule S'

1 Generate possible priority assignment array *pos_priority*[] by *priority_adj*(λ_{new});
2 **while** *pos_priority*[] *is not empty* **do**
3     $\lambda_{fine} \leftarrow$ *pos_priority*[*i*];
4   *LPES*(λ_{fine});
5   **if** *solution_feasible and* ($E_{fine} < E_{opt}$) **then**
6 | $\lambda_{opt} \leftarrow \lambda_{fine}$, $E_{opt} \leftarrow E_{fine}$;
7   **end**
8 **end**

In Algorithm 2, firstly, the priority assignments of tasks on each core are recorded according to the accepted solution λ_{new}. Then, the FT keeps strictly periodic tasks unchanged. For tie-breaking and non-strictly periodic tasks, it adjusts and records their priorities in possible priority assignment array *pos_priority*[] by using *priority_adj*() function. Next, for each element in *pos_priority*[], the algorithm performs *LPES*. In each iteration, the feasible solution λ_{fine} (checked for feasibility by *solution_feasible* function) that can reduce the energy consumption is stored. Finally, the tasks are adjusted iteratively until no improvement can be achieved. Note that FT can be used directly in the optimization process to find an optimal solution quickly if the initial solution is good. The FT scheme is illustrated through the following example.

Example 2. *In Example 1, the initial execution order of tasks on CORE1 and CORE2 are* $\{v_{7,1}^1 \to v_{6,1}^1 \to v_{7,2}^1 \to v_{2,1}^1 \to v_{6,2}^1 \to v_{4,1}^1\}$ *and* $\{v_{5,1}^2 \to v_{8,1}^2 \to v_{1,1}^2 \to v_{3,1}^2 \to v_{5,2}^2 \to v_{8,2}^2\}$, *respectively. Among them, task* $v_{2,1}^1$ *and task* $v_{6,2}^1$ *are tie breaking tasks, and tasks belonging to application* g_2 *are non-strictly periodic. Thus, they can be adjusted (swapped) as long as the precedence constraints are guaranteed. The corresponding schedule after FT can be seen in Figure 5b, execution order of tasks on CORE1 and CORE2 are* $\{v_{7,1}^1 \to v_{6,1}^1 \to v_{7,2}^1 \to v_{6,2}^1 \to v_{2,1}^1 \to v_{4,1}^1\}$ *and* $\{v_{1,1}^2 \to v_{3,1}^2 \to v_{5,1}^2 \to v_{8,1}^2 \to v_{5,2}^2 \to v_{8,2}^2\}$, *respectively. The improvement of power consumption on the system after FT is, therefore, 8.2%.*

6. Experiment Evaluation

This section presents the experimental setup and case studies. To evaluate and demonstrate the efficiency of our proposed approaches, the experiments are performed on a 3.60 GHz 4-core PC with 4 GB of memory under Windows 7. The same 70 nm technology power parameters of the processor are used as in the studies [17,42,44,45]. Code is written in C language, and we use the IBM ILOG CPLEX 12.5.0.0 Solver to solve the MILP formulations. In each case, CPLEX is given a time limit of 10 h.

6.1. Experiment Setup

Our experiments include 12 applications represented by task graph TG1–TG12. TG1–TG3 are based on industrial, automotive, and consumer applications [47]. TG4–TG6 are three applications from 'Standard Task Graph' [48], which are based on real applications, namely, a robotic control application, the FPPPP SPEC benchmark and a sparse matrix solver. In addition, we use a general randomized task-graph generator TGFF [49] to create six different periodic applications (TG7–TG12) with typical industrial characteristics in our experiments. These task graphs are from the original example input file (e.g., kbasic, kseries-paralle and robtst) that come from the software package. Then, we consider nine combinations (i.e., five relatively small benchmarks, namely SG1–SG5, and four large benchmarks, namely LG1–LG4) of these task graphs from TG1–TG12. Each benchmark has 2–5 task graphs with features including different topologies (such as chain, in-tree, out-tree, and fork-join), different lengths of critical paths and numbers of dependent tasks. The period of the task graphs are distributed randomly in [10, 2000] ms. We define a parameter α varied from [0, 1] for the whole set of tasks in each benchmark, which reflects the ratio between the strictly and non-strictly periodic tasks. In other words, all tasks are strictly periodic if α is equal to 1, and non-strictly periodic if α is equal 0.

We consider a 4-core architecture for our experiment. The power model is based on a typical 70 nm technology processor, which has been applied in the works [17,41,42,44,45]. The accuracy of the processor power model has been verified by SPICE simulation. For fairness of comparison, parameters of cores power, voltage levels and energy overhead of processor mode switch are referred to [17]. As shown in Table 4, the processor can operate at five voltage levels within the range of 0.65 V to 0.85 V with 50 mV steps and the corresponding frequencies vary from 1.01 GHz to 2.10 GHz. The corresponding dynamic power P_d and static power P_s under different v/f level are calculated according to the energy model in Section 3.3 and the technology constants (e.g., C_{ef}, k, and V_t) from [42,44,45]. The time overhead of processor mode switch t_{ms} and voltage/frequency switch are 10 ms and 600 μs, respectively, from [50]. For the mapping step, we use the task assignment algorithm in [31] to assign each task to the MPSoC.

Table 4. Core power model parameters.

Level	1	2	3	4	5
V_{dd} (V)	0.85	0.8	0.75	0.7	0.65
f (GHz)	2.10	1.81	1.53	1.26	1.01
P_d (mW)	655.5	489.9	370.4	266.7	184.9
P_s (mW)	462.7	397.6	340.3	290.1	246
	P_{idle} = 276 mW;	P_{sleep} = 0.08 mW;	E_{ms} = 0.385 mJ		

6.2. Experiment Results

This section presents the evaluation of our improved MILP method in Section 5.1 and heuristic algorithm in Section 5.2. The number of tasks and edges of each benchmark is shown in first column of Table 5.

Table 5. Average power consumption under different methods.

Benchmarks (Tasks/Edges)	Power Consumption (W)						
	SMILP	IMILP			HLTSA-FT		
	$\forall \alpha$	$\alpha = 3/4$	$\alpha = 1/2$	$\alpha = 1/4$	$\alpha = 3/4$	$\alpha = 1/2$	$\alpha = 1/4$
SG1 (13/15)	2.71	2.49	2.32	2.15	2.49	2.33	2.15
SG2 (19/16)	2.96	2.61	2.39	2.14	2.63	2.39	2.17
SG3 (22/12)	2.73	2.57	2.42	2.21	2.58	2.45	2.24
SG4 (25/31)	2.98	2.74	2.55	2.48	2.76	2.56	2.49
SG5 (34/23)	4.11	3.78	3.60	3.49	3.82	3.63	3.51

Table 5. *Cont.*

Benchmarks (Tasks/Edges)	Power Consumption (W)						
	SMILP	IMILP			HLTSA-FT		
	$\forall\alpha$	$\alpha = 3/4$	$\alpha = 1/2$	$\alpha = 1/4$	$\alpha = 3/4$	$\alpha = 1/2$	$\alpha = 1/4$
Average Reduction	-	8.38%	14.38%	19.86%	7.92%	13.88%	19.48%
			14.21%			13.76%	
LG1 (108/72)	TL	TL	TL	TL	3.93	3.41	3.37
LG2 (230/241)	TL	TL	TL	TL	4.89	4.26	3.79
LG3 (355/329)	TL	TL	TL	TL	4.77	4.39	3.55
LG4 (416/263)	TL	TL	TL	TL	4.11	3.70	3.14

6.2.1. Evaluation of the Improved MILP

We evaluate and compare our improved MILP method (represented by IMILP) with existing scheduling method (denoted as SMILP) in which the periodic constraint must be strictly followed in the start of all tasks [17]. Table 5 shows average power consumptions in one hyper-period (i.e., the average value of $E_{total_H_a}$ divided by H_a) under different MILP-based methods. The results are obtained in three different cases with the factor α varying from 1/4 to 3/4 with step size 1/4.

From Table 5, one can see that SMILP fails to increase energy savings in contrast to our IMILP. Compared with SMILP, the IMILP in case α = 3/4, 1/2 and 1/4 reduces power consumption for small benchmarks SG1-SG5 by, 8.38%, 14.38% and 19.86%, respectively. The average power consumption can be reduced by 14.21%. The results demonstrate that the simplistic assumption in previous SMILP methods where each task and its task instances must strictly start periodically can lead to an increase in energy consumption. On the other hand, column "SMILP" under "Power Consumption (W)" illustrates the power consumption under SMILP in any cases of α remain unchanged. However, the results under IMILP (from column 3–5 and 6–8) show that the smaller value of α, the more power consumption can be reduced. This is because our IMILP can capture the periodicity of specific tasks belonging to their applications, and deals with strictly and non-strictly periodic tasks correctly. To exploit energy-savings, the IMILP method effectively utilizes the flexibility of non-strictly periodic tasks in scheduling as the tasks do not need to start periodically.

6.2.2. Evaluation of the HLTSA-FT

We first compare HLTSA-FT heuristic method with MILP-based methods. The average power consumptions under different α are listed in the last three columns in Table 5. Compared with SMILP, the HLTSA-FT in case α = 3/4, 1/2 and 1/4 reduces power consumption for small benchmarks SG1–SG5 by 7.92%, 13.88% and 19.48%, respectively. The power consumption can be reduced on average, by 13.76%. For the five test cases, the average (minimum) deviation of the HLTSA-FT from the IMILP is only 3.2% (1.9%). The result demonstrates our HLTSA-FT heuristic can find near optimal solutions and its performance is close to that of IMILP for SG1–SG5. Although the MILP method can obtain optimal results, the computation time of the method grows exponentially with increasing size of benchmarks as shown in columns 2–4 in Table 6. The sign 'TL' in Tables 5 and 6 indicates that the MILP methods for LG1–LG4 cannot generate any optimal solution in limited time (10 h in our experiment). This verifies that the ILP solver fails to find the optimal solutions for models with large instances. However, our HLTSA-FT heuristic algorithm can always generate feasible solutions efficiently for the large benchmarks LG1–LG4. Thus, HLTSA-FT provides a good way for designers to search for energy-efficient scheduling when computation time is intolerable.

Table 6. Average computation time under different methods for various α.

Benchmarks	IMILP			LSA			HLTSA			HLTSA-FT		
	$\alpha = 3/4$	$\alpha = 1/2$	$\alpha = 1/4$	$\alpha = 3/4$	$\alpha = 1/2$	$\alpha = 1/4$	$\alpha = 3/4$	$\alpha = 1/2$	$\alpha = 1/4$	$\alpha = 3/4$	$\alpha = 1/2$	$\alpha = 1/4$
SG1	8.115	8.856	12.838	4.634	4.851	5.193	0.983	1.137	1.602	1.204	1.252	2.37
SG2	12.306	18.974	21.244	4.315	9.183	15.589	4.902	5.926	10.39	4.991	6.878	9.649
SG3	22.928	32.730	96.795	20.059	28.308	31.443	7.998	10.341	30.464	8.656	9.908	32.004
SG4	65.983	97.253	112.581	39.229	50.943	112.556	15.831	16.955	28.77	18.534	18.775	27.639
SG5	624.600	1839.851	2603.3	178.315	205.233	255.689	27.041	27.541	51.837	27.426	28.53	55.51
LG1	TL	TL	TL	846.9	2899.5	5220.98	504.94	1022.7	2455.3	632.06	1349.56	1904.56
LG2	TL	TL	TL	594.88	4534.12	10,120.82	123.18	2611.54	2956.88	318.8	3286.26	4294.3
LG3	TL	TL	TL	3695.8	24,170.1	34,201.16	920.36	2329.4	3645.92	849.62	2361.98	3283.77
LG4	TL	TL	TL	9639.8	23,339.78	35,815.02	2858.25	2903.53	5439.96	3553.46	4513.72	6601.95

We then compare HLTSA-FT with existing heuristic algorithms [38–40]. As mentioned in Section 2, the SA-based algorithms have been widely applied to achieve near-optimal solutions for low power scheduling. For fair comparison, we modify and implement the energy-efficient scheduling methods for our problem under three configurations: LSA, HLTSA, and HLTSA-FT. The LSA heuristic applies the list-based SA algorithm but does not consider tabu list and fine-tune phase. The HLTSA heuristic considers LSA integrating tabu list but no fine-tune phase. We evaluate and compare our HLTSA-FT algorithm with these heuristics in terms of two performance metrics: (1) the solution quality and (2) the computation time of searching process.

One can see that the HLTSA and HLTSA-FT outperform LSA in solution quality. The comparison of average power consumption of HLTSA-FT with those of SA-based algorithms are presented in Figures 7 and 8, respectively. In Figure 7, for small benchmarks SG1–SG5, the HLTSA and HLTSA-FT reduce average power consumption by 4.41% and 13.31%, respectively, compared with LSA. In Figure 8, for large benchmarks LG1–LG4, the HLTSA and HLTSA-FT reduce average power consumption by 6.09% and 19.27%, respectively, compared with LSA. This is due to the fact that HLTSA and HLTSA-FT use short-term memory of recently visited solutions known as tabu list in SA to escape from local optima. The search can be restricted from retiring to a previously visited solution and performance of SA can be enhanced significantly with help of the tabu list.

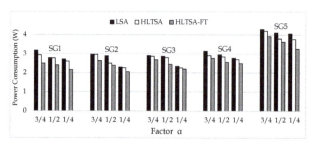

Figure 7. Comparison of average power consumption under different heuristics (SG1–SG5).

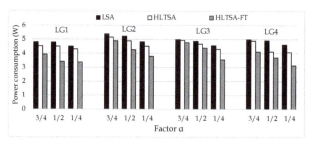

Figure 8. Comparison of average power consumption under different heuristics (LG1–LG4).

Moreover, our HLTSA-FT improves the solution quality in contrast to HLTSA. In Figure 7, for small benchmarks SG1–SG5, the HLTSA-FT in case $\alpha = 3/4, 1/2$ and $1/4$ reduces average power consumption by 8.79%, 8.91% and 10.11%, respectively, compared with HLTSA. In Figure 8, for large benchmarks LG1-LG4, the HLTSA-FT in case $\alpha = 3/4, 1/2$ and $1/4$ reduces average power consumption by 9.37%, 13.32% and 19.61%, respectively, compared with HLTSA. The reason lies in the fact that the performance is significantly improved by introducing the FT phase. The HLTSA without FT phase focuses on searching for better (concerning energy) solutions blindly and randomly, a lot of which are however abandoned because of violation of corresponding precedence and deadline constraints. The HLTSA-FT actively looks around for near solutions and leads the way to potential energy-efficient schedules by adjusting execution order of tasks if the precedence and deadline constraints are satisfied.

The column 5–13 in Table 6 presents average computation time under different SA-based methods for various α. The comparison results are obtained over 10 runs when solving our problem. As can be seen from the table, an interesting observation is that the computation time increases as α decreases. This is caused by the fact that, as α decreases, the number of constraints for specific strictly periodic tasks decreases and the search space of the problem becomes larger. This just demonstrates that our problem requires an effective heuristic algorithm to reduce complexity when the input size becomes larger.

HLTSA and HLTSA-FT outperform LSA on the convergence speed. For example, compared with LSA for different benchmarks (SG1–SG5, LG1–LG4), the HLTSA algorithm under $\alpha = 1/2$ reduces the average computation time by 76.6%, 35.5%, 63.5%, 66.7%, 86.6%, 64.7%, 42.4%, 90.4%, and 87.6%, respectively, and the HLTSA-FT algorithm under $\alpha = 1/2$ reduces the average computation time by 74.2%, 25.1%, 65.0%, 63.1%, 86.1%, 53.5%, 27.5%, 90.2% and 80.7%, respectively. This is because LSA does not keep track of recently visited solutions and needs more iterations to find the best solution, while HLTSA and HLTSA-FT exploit the beauty of tabu search and simulated annealing to ensure the convergence at faster rate. Furthermore, one can observe that our HLTSA-FT algorithm can improve the solution quality (in Figures 7 and 8), without increasing significantly the number of required simulations compared with the HLTSA (in Table 6).

To summarize, the experimental results presented above show that the proposed HLTSA-FT heuristic algorithm achieves a good trade-off between solution quality and solution generation time compared with the IMILP, LSA and HLTSA methods as the problem scale becomes larger. The algorithm is a scalable heuristic method that users can adjust the configuration parameters of the algorithm according to the specific input. To achieve further performance improvement, the HLTSA-FT algorithm can obtain high-quality solutions by increasing optimization iterations or executing multiple times within an acceptable time.

7. Conclusions

This paper has investigated the problem of scheduling a set of periodic applications for energy optimization on safety/time-critical time-triggered systems. In the applications, besides strictly periodic tasks, there also exist non-strictly periodic tasks in which different invocations of a task can start aperiodically. We present a practical task model to characterize the strictness of the task's periodicity, and formulate a novel scheduling problem for energy optimization based on the model. To address the problem, we first propose an improved MILP model to obtain energy-efficient scheduling. Although the MILP method can generate optimal solutions, its solution computation time grows exponentially with the number of inputs. Therefore, we further develop an HLTSA-FT algorithm to reduce complexity and efficiently obtain a high-quality solution within a reasonable time. Extensive evaluations on both synthetic and realistic benchmarks have demonstrated the effectiveness of our improved MILP method and the HLTSA-FT algorithm, compared with the existing studies.

Some issues are taken into account in our future work. In this paper, we assume that task mappings are given as a fixed input. For a higher energy efficiency, mapping and scheduling on time-triggered multiprocessor systems need to be integrated since they are inter-dependent. Currently, we are working on solving this problem. Furthermore, we intend to study how to integrate our approaches with online scheduling methods on the realistic safety/time-critical multiprocessor systems to leverage system-wide energy consumption.

Author Contributions: X.J. conceived and developed the ideas behind the research, performed the experiments and wrote the paper under the supervision of K.H. Authors K.H., X.Z., R.Y., K.W. and D.X. provided guidance and key suggestions. K.H. and X.Y. supervised the research and finalized the paper.

Funding: This research was funded by National Science and Technology Major Project grant number 2017ZX01030-102-002.

Acknowledgments: This research was supported by the National Science and Technology Major Project under Grant 2017ZX01030-102-002.

Conflicts of Interest: The authors declare no conflict of interest.

References

1. Kang, S.H.; Kang, D.; Yang, H.; Ha, S. Real-time co-scheduling of multiple dataflow graphs on multi-processor systems. In Proceedings of the 2016 Design Automation Conference, Austin, TX, USA, 5–9 June 2016; pp. 1–6.
2. García-Valls, M.; Cucinotta, T.; Lu, C. Challenges in real-time virtualization and predictable cloud computing. *J. Syst. Archit.* **2014**, *60*, 726–740. [CrossRef]
3. Kopetz, H. The Time-Triggered Model of Computation. In Proceedings of the IEEE Real-Time Systems Symposium, Madrid, Spain, 4 December 1998; Volume 2, pp. 168–177.
4. Mittal, S. A Survey of Techniques for Improving Energy Efficiency in Embedded Computing Systems. *Int. J. Comput. Aided Eng. Technol.* **2014**, *6*, 440–459. [CrossRef]
5. Zhang, Y.; Hu, X.; Chen, D.Z. Task Scheduling and Voltage Selection for Energy Minimization. In Proceedings of the Design Automation Conference, New Orleans, LA, USA, 10–14 June 2002; pp. 183–188.
6. Baskiyar, S.; Abdel-Kader, R. Energy aware DAG scheduling on heterogeneous systems. *Cluster Comput.* **2010**, *13*, 373–383. [CrossRef]
7. Luo, J.; Jha, N.K. Static and Dynamic Variable Voltage Scheduling Algorithms for Real-Time Heterogeneous Distributed Embedded Systems. In Proceedings of the IEEE International Conference on VLSI Design, Bangalore, India, 11 January 2002; p. 719.
8. Liu, Y.; Veeravalli, B.; Viswanathan, S. Novel critical-path based low-energy scheduling algorithms for heterogeneous multiprocessor real-time embedded systems. In Proceedings of the International Conference on Parallel and Distributed Systems, Hsinchu, Taiwan, 5–7 December 2007; pp. 1–8.
9. Kanoun, K.; Mastronarde, N.; Atienza, D.; Van der Schaar, M. Online Energy-Efficient Task-Graph Scheduling for Multicore Platforms. *IEEE Trans. Comput.-Aided Des. Integr. Circuits Syst.* **2014**, *33*, 1194–1207. [CrossRef]
10. Kianzad, V.; Bhattacharyya, S.S.; Qu, G. CASPER: An Integrated Energy-Driven Approach for Task Graph Scheduling on Distributed Embedded Systems. In Proceedings of the IEEE International Conference on Application-Specific Systems, Architecture Processors, Samos, Greece, 23–25 July 2005; pp. 191–197.
11. Zhou, J.; Yan, J.; Cao, K.; Tan, Y.; Wei, T.; Chen, M.; Zhang, G.; Chen, X.; Hu, S. Thermal-Aware Correlated Two-Level Scheduling of Real-Time Tasks with Reduced Processor Energy on Heterogeneous MPSoCs. *J. Syst. Archit.* **2017**, *82*, 1–11. [CrossRef]
12. Davare, A.; Zhu, Q.; Natale, M.D.; Pinello, C.; Kanajan, S.; Vincentelli, A.S. Period optimization for hard real-time distributed automotive systems. In Proceeding of the 44th annual Design Automation Conference, San Diego, CA, USA, 4–8 June 2007; pp. 278–283.
13. Lukasiewycz, M.; Schneider, R.; Goswami, D.; Chakraborty, S. Modular scheduling of distributed heterogeneous time-triggered automotive systems. In Proceeding of the Design Automation Conference, Sydney, NSW, Australia, 30 January–2 February 2012; pp. 665–670.
14. Balogh, A.; Pataricza, A.; Rácz, J. Scheduling of Embedded Time-triggered Systems. In Proceedings of the 2007 Workshop on Engineering Fault Tolerant Systems (EFTS '07), Dubrovnik, Croatia, 4 September 2007; ACM: New York, NY, USA, 2007.
15. Chen, J.J.; Schranzhofer, A.; Thiele, L. Energy minimization for periodic real-time tasks on heterogeneous processing units. In Proceedings of the IEEE International Symposium on Parallel & Distributed Processing, Rome, Italy, 23–29 May 2009; pp. 1–12.

16. Pop, P. Scheduling and voltage scaling for energy/reliability trade-offs in fault-tolerant time-triggered embedded systems. In Proceedings of the IEEE/ACM International Conference on Hardware/Software Codesign and System Synthesis, Salzburg, Austria, 30 September–3 October 2007; pp. 233–238.

17. Chen, G.; Huang, K.; Knoll, A. Energy optimization for real-time multiprocessor system-on-chip with optimal DVFS and DPM combination. *ACM Trans. Embed. Comput. Syst.* **2014**, *13*, 111. [CrossRef]

18. Sagstetter, F.; Lukasiewycz, M.; Chakraborty, S. Generalized Asynchronous Time-Triggered Scheduling for FlexRay. *IEEE Trans. Comput.-Aided Des. Integr. Circuits Syst.* **2017**, *36*, 214–226. [CrossRef]

19. Freier, M.; Chen, J.J. Time Triggered Scheduling Analysis for Real-Time Applications on Multicore Platforms. In Proceedings of the RTSS Workshop on REACTION, Rome, Italy, 2–5 December 2014; pp. 43–52.

20. Gendy, A.K.G. Techniques for Scheduling Time-Triggered Resource-Constrained Embedded Systems. University of Leicester, Leicester, UK, 2009.

21. Hu, M.; Luo, J.; Wang, Y.; Veeravalli, B. Scheduling periodic task graphs for safety-critical time-triggered avionic systems. *IEEE Trans. Aerosp. Electron. Syst.* **2015**, *51*, 2294–2304. [CrossRef]

22. Freier, M.; Chen, J.J. Sporadic Task Handling in Time-Triggered Systems. In Proceedings of the International Workshop on Software and Compilers for Embedded Systems, Sankt Goar, Germany, 23–25 May 2016; pp. 135–144.

23. Kramer, S.; Ziegenbein, D.; Hamann, A. Real world automotive benchmark for free. In Proceedings of the International Workshop on Analysis Tools and Methodologies for Embedded and Real-Time Systems, Lund, Sweden, 7 July 2015.

24. García-Valls, M.; Domínguez-Poblete, J.; Touahria, I.E.; Lu, C. Integration of Data Distribution Service and distributed partitioned systems. *J. Syst. Archit.* **2017**, *83*, 23–31. [CrossRef]

25. García-Valls, M.; Calva-Urrego, C. Improving service time with a multicore aware middleware. In Proceedings of the Symposium on Applied Computing, Marrakech, Morocco, 3–7 April 2017; pp. 1548–1553.

26. Qiu, K.; Gong, Z.; Zhou, D.; Chen, W.; Xu, Y.; Shi, X.; Liu, Y. Efficient Energy Management by Exploiting Retention State for Self-powered Nonvolatile Processors. *J. Syst. Archit.* **2018**, *87*, 22–35. [CrossRef]

27. Ghadaksaz, E.; Safari, S. Storage Capacity for EDF-ASAP Algorithm in Energy-Harvesting Systems with Periodic Implicit Deadline Hard Real-Time Tasks. *J. Syst. Archit.* **2018**. [CrossRef]

28. Devadas, V.; Aydin, H. On the Interplay of Voltage/Frequency Scaling and Device Power Management for Frame-Based Real-Time Embedded Applications. *IEEE Trans. Comput* **2011**, *61*, 31–44. [CrossRef]

29. Gerards, M.E.T.; Kuper, J. Optimal DPM and DVFS for Frame-based Real-time Systems. *ACM Trans. Archit. Code Optim.* **2013**, *9*, 41. [CrossRef]

30. Kwok, Y.K.; Ahmad, I. Static scheduling algorithms for allocating directed task graphs to multiprocessors. *ACM Comput. Surv.* **1999**, *31*, 406–471. [CrossRef]

31. Schmitz, M.; Al-Hashimi, B.; Eles, P. Energy-Efficient Mapping and Scheduling for DVS Enabled Distributed Embedded Systems. In Proceedings of the Conference on Design, Automation and Test in Europe (DATE '02), Paris, France, 4–8 March 2002; pp. 514–521.

32. Guzek, M.; Pecero, J.E.; Dorronsoro, B.; Bouvry, P. Multi-objective evolutionary algorithms for energy-aware scheduling on distributed computing systems. *Appl. Soft Comput. J.* **2014**, *24*, 432–446. [CrossRef]

33. Xie, G.; Zeng, G.; Liu, L.; Li, R.; Li, K. Mixed real-time scheduling of multiple DAGs-based applications on heterogeneous multi-core processors. *Microprocess. Microsyst.* **2016**, *47*, 93–103. [CrossRef]

34. Xie, G.; Zeng, G.; Liu, L.; Li, R.; Li, K. High performance real-time scheduling of multiple mixed-criticality functions in heterogeneous distributed embedded systems. *J. Syst. Archit.* **2016**, *70*, 3–14. [CrossRef]

35. Zhang, C.Y.; Li, P.G.; Rao, Y.Q.; Guan, Z.L. A very fast TS/SA algorithm for the job shop scheduling problem. *Comput. Oper. Res.* **2008**, *35*, 282–294. [CrossRef]

36. Katsigiannis, Y.A.; Georgilakis, P.S.; Karapidakis, E.S. Hybrid Simulated Annealing-Tabu Search Method for Optimal Sizing of Autonomous Power Systems With Renewables. *IEEE Trans. Sustain. Energy* **2012**, *3*, 330–338. [CrossRef]

37. Chan, F.T.; Prakash, A.; Ma, H.; Wong, C. A hybrid Tabu sample-sort simulated annealing approach for solving distributed scheduling problem. *Int. J. Prod. Res.* **2013**, *51*, 2602–2619. [CrossRef]

38. He, D.; Mueller, W. A Heuristic Energy-Aware Approach for Hard Real-Time Systems on Multi-core Platforms. In Proceedings of the Euromicro Conference on Digital System Design, Izmir, Turkey, 5–8 September 2012; pp. 288–295.

39. Luo, J.; Jha, N.K. Power-Efficient Scheduling for Heterogeneous Distributed Real-Time Embedded Systems. *IEEE Trans. Comput.-Aided Des. Integr. Circuits Syst.* **2007**, *26*, 1161–1170. [CrossRef]

40. Ni, J.; Wang, N.; Yoshimura, T. Tabu search based multiple voltage scheduling under both timing and resource constraints. In Proceedings of the International Symposium on Quality Electronic Design, Santa Clara, CA, USA, 2–4 March 2015; pp. 118–122.

41. Wang, Y.; Liu, H.; Liu, D.; Qin, Z.; Shao, Z.; Sha, H.M. Overhead-aware energy optimization for real-time streaming applications on multiprocessor System-on-Chip. *ACM Trans. Des. Autom. Electron. Syst.* **2011**, *16*, 14. [CrossRef]

42. Wang, W.; Mishra, P. Leakage-Aware Energy Minimization Using Dynamic Voltage Scaling and Cache Reconfiguration in Real-Time Systems. In Proceedings of the International Conference on VLSI Design, Bangalore, India, 3–7 January 2010; pp. 357–362.

43. Peng, D.T.; Shin, K.G.; Abdelzaher, T.F. Assignment and scheduling communicating periodic tasks in distributed real-time systems. *IEEE Trans. Softw. Eng.* **1997**, *23*, 745–758. [CrossRef]

44. Martin, S.M.; Flautner, K.; Mudge, T.; Blaauw, D. Combined dynamic voltage scaling and adaptive body biasing for lower power microprocessors under dynamic workloads. In Proceedings of the IEEE/ACM International Conference on Computer-Aided Design, San Jose, CA, USA, 10–14 November 2002; pp. 721–725.

45. Jejurikar, R.; Pereira, C.; Gupta, R. Leakage aware dynamic voltage scaling for real-time embedded systems. In Proceedings of the Design Automation Conference, San Diego, CA, USA, 7–11 June 2004; pp. 275–280.

46. Coskun, A.K.; Whisnant, K.A.; Gross, K.C. Static and dynamic temperature-aware scheduling for multiprocessor SoCs. *IEEE Trans. Very Large Scale Integr. Syst.* **2008**, *16*, 1127–1140. [CrossRef]

47. Embedded Microprocessor Benchmark Consortium. Available online: http://www.eembc.org/ (accessed on 18 June 2018).

48. Tobita, T.; Kasahara, H. A standard task graph set for fair evaluation of multiprocessor scheduling algorithms. *J. Sched.* **2002**, *5*, 379–394. [CrossRef]

49. Dick, R.P.; Rhodes, D.L.; Wolf, W. TGFF: Task graphs for free. In Proceedings of the Sixth International Workshop on Hardware/Software Codesign (CODES/CASHE '98), Seattle, WA, USA, 18 March 1998; pp. 97–101.

50. *Marvell PXA270 Processor Electrical, Mechanical, Thermal Specification*; Marvell: Hamilton, Bermuda, 2009.